MICHAEL PATRICK F. SMght currently based in central ....dy Guthrie Dreams and Ain't N re and New York. As a musician .....olk luminaries such as Ramblin' ..... Elliott. He has also worked as a stage actor, junk hauler, furniture mover, legal assistant, book store clerk, contractor, driver, office temp, stagehand, set fabricator, bartender, and now writer. *The Good Hand* is his first book.

Praise for *The Good Hand*:

'After reading *The Good Hand* you may reassess whether you have ever truly done a hard day's work in your life ... This lyrical and engrossing memoir is an extraordinary tale ... Smith writes movingly of his chaotic childhood ... the tragedies slowly drip out ... There have been predictable comparisons to other recent hardship autobiographies – JD Vance's *Hillbilly Elegy* and Tara Westover's *Educated* – but Smith's story, blessedly, comes with more (crude) humour ... Undeniably powerful'   *Sunday Times*

'[A] sprawling, heart-smeared-on-the-page howl of rage and pain. *The Good Hand* is a rambling honky-tonk of a book, with the soul of a songwriter and the ache of a poor white boy who grew up rough. It is big and it is pretty and it is amazing' *Los Angeles Times*

'Beautiful, funny and harrowing ... As someone whose immediate family bears the scars of physical labour in another Great Plains state, and who rarely sees her native class convincingly portrayed, I relished these anecdotes and the validation they provide'
                                                                    The Atlantic

'Smith brings an alchemic talent to describing physical labour ... With a playwright's talent for dialogue, storytelling in miniature and staying out of the way, he writes dozens of scenes of men moving, joking and endlessly talking ... his writing keeps people alive in their histories, talents, humour and mistakes ... [bringing] perspective, on how people, including Smith, can sometimes rise above their worst selves through unglamorous, demanding, difficult work ... a book that should be read'   *New York Times Book Review*

'Smith guides us through a long muddy year in North Dakota's oil boom, telling the story as if it were a Sam Shepard play that can't be stopped even though people are about to get badly hurt. It's a surprisingly tender account of a man who is searching for salvation – from the sins of his family, from the drunken and drugged-up sins of a world broken by corporations – while trying desperately to find himself through work'

ROBERT SULLIVAN, author of *The Thoreau You Don't Know*

'A sincere and colourful account of down-and-out men trying to make it and maybe grow up in the eternal dreary tailgate party and crushing dangerous toil of the fracking boom. As one of Smith's mentors tells him, "Now you know why gas is so expensive"'

WILLIAM T. VOLLMANN, author of *The Lucky Star*

'A fine successor to Orwell's *Down and Out in Paris and London*, translated to the epic scope and sweep of the prairie West. While telling his bruisingly candid tales of toiling in a North Dakota oil field, Smith paints a larger portrait of the American worker's heroically tragic struggle for a thin slice of the pie'

JOHN STRAUSBAUGH, author of *Victory City*

THE GOOD HAND

—

A MEMOIR OF WORK,

BROTHERHOOD AND

TRANSFORMATION IN AN

AMERICAN BOOMTOWN

—

MICHAEL PATRICK F. SMITH

WILLIAM
COLLINS

William Collins
An imprint of HarperCollins*Publishers*
1 London Bridge Street
London SE1 9GF

WilliamCollinsBooks.com

HarperCollins*Publishers*
1st Floor, Watermarque Building, Ringsend Road
Dublin 4, Ireland

First published in Great Britain in 2021 by William Collins
This William Collins paperback edition published in 2022

1

A catalogue record for this book is
available from the British Library

ISBN 978-0-00-839948-1

Set in Berling LT Std
Printed and bound in the UK using 100%
renewable electricity at CPI Group (UK) Ltd

MIX
Paper from
responsible sources
FSC™ C007454
www.fsc.org

For my brothers and sisters:

Ryan, Matthew, Megan, Kate, and Shanon

The world is charged with the grandeur of God.
It will flame out, like shining from shook foil;
It gathers to a greatness, like the ooze of oil
Crushed. Why do men then now not reck his rod?
Generations have trod, have trod, have trod;
And all is seared with trade; bleared, smeared with toil;
And wears man's smudge and shares man's smell: the soil
Is bare now, nor can foot feel, being shod.

—GERARD MANLEY HOPKINS

RAISING THE DERRICK

Like most people who rely on oil for nearly every aspect of their existence, I didn't know anything about a drilling rig. I pulled off my hard hat, wiped my forehead with my forearm, and looked across the work location, trying to take it all in. On a mud-colored scar in the land cut flat two hundred yards across, trucks coughed diesel, cranes farted smoke, and dust turned to mud on the faces of men toiling under bright sunlight. The rig was half in pieces—giant square chunks of steel, littered across the dirt. I had no idea how it went back together. I adjusted the inside strap of my hard hat—what had the roughneck called it? A "brain bucket?"—and set it back on my head. It was my fifth day in the oil patch. I had yet to make a hand; I didn't even know what it meant to "make a hand," and I had no idea how vital that idea would become to me. I was a greenhorn, what most oil patch guys call a "worm," a headless, legless, brainless thing, good for eating dirt and shutting the fuck up.

I had just left the shit box. The plumbing wasn't functional yet, so the bathroom consisted of a trailer with a bare lightbulb, a toilet, and a mess of flies. It was 99 degrees outside the trailer and probably 120 degrees inside the shit box. Standing twenty feet away from it, I could smell the stink carried on the breeze.

In five days of work, I'd clocked over seventy hours of heavy manual labor. My neck and shoulders were stiff, my arms, legs, knees, and ankles battered and sore. My hands ached, and my head felt foggy from lack of sleep. Over the past several weeks, I had been crashing on a filthy flophouse floor, in a room I shared with several other migrant workers. The stress of the oil field work, combined with the stress of this living situation, aggravated a twitch under my left eye. It felt like my face was shorting out.

That morning, I'd made the mistake of thinking about giving up. The thought had wormed its way into my brain, and now I couldn't seem to shake it. I was bone-tired, afraid, and nauseous.

There were many ways to get hurt on location. Or die. Exhausted men operated heavy machinery, trucks thundered past, heavy objects flew through the air—steel swinging steel around on cables, hooks, and slings. You could go blind. You could lose your hearing. You could lose a limb. A rod could snap, and bam: instant lobotomy. One hundred thirty-eight oil field hands had been killed in the United States the previous year. I looked around. I could see it happening to me. My breathing sped up, and my stomach knotted. I tried to take slow steady breaths.

Dragonflies buzzed around. I walked through a row of pickup trucks, across the dirt patch and toward a pile of wind walls—thin metal sheets some fifteen feet across. A wind wall does exactly what it sounds like it should do. On the high plains, with no trees or mountains to stop it, the wind is fierce. Even in the summer, gusts of northern air can all but knock you down. The walls protect against that. They make it easier for men to work the floor of the drilling rig, two to three stories above the dirt.

I was twenty yards from the crane and maybe thirty yards from the base of the rig. I squinted up into the sun. A wind wall, attached to the crane by four metal chains on the crane's whip line, dangled in the air above the rig's floor. A crew of roughnecks, strapped into harnesses and wearing heavy work boots, guided the wall down, set it carefully in place, and, after lumbering along the edge of the floor, swung their sledgehammers at large metal pins to attach it to the rig.

When the wall was connected, the crane operator reeled in the whip line. It swung in an arc above the men, away from them and toward me. With the precision of a bullet hitting a bull's-eye, the chains came to a sharp stop directly in front of me, the hooks suspended at the level of my chest. I gathered them into my hands, sliding the fasteners into eye hooks welded onto the corners of the next wall, then I looked back to the operator. He sat in the cab, blue Oakley sunglasses and beard slapped across an inscrutable expression. I waved a hand, and he gave me a nod. I stepped away from the pile, and another wind wall flew up into the blue.

AT LUNCHTIME, I sat in the back of the work van and ate cold Chunky soup out of the can. Bobby Lee sat with the driver's seat kicked way back, his boots up on the dash. He wore a Resistol brand Diamond Horseshoe cowboy hat pulled low over his eyes. At one point the hat had been the color of pearl, but it was beat to shit, dirty, greasy, and floppy—incongruent with his studied look. "Now you know why gas is so expensive," Bobby Lee said.

I stared out the window of the van. The work site was cluttered with tractor trailers, pickup trucks, forklifts, a hydraulic crane, a lattice boom crane, rows of stacked piping, giant metal structures, and crews of men.

No one was even drilling for oil yet. Everything I saw out the window—the trucks, the cranes, the men—was there because a previous well had been drilled to completion. And so we had shown up, Diamondback Trucking, to take the rig—the mechanism that extracts oil from the well—apart and load it onto haul trucks to bring it here, to a new location, where it was being reassembled over a new drill hole.

The previous day, Bobby Lee had handed me a sheet of paper to sign. In the mornings, we had safety meetings. We stood around with the roughnecks; somebody talked over the day's plans, and we all signed a paper. I had never read the paper, but when Bobby handed me this page I thought it looked like all the others.

"Didn't I sign this already?" I asked him.

"It's a new one," Bobby said.

"Okay," I said, putting my name across it.

"And when I hand you something," Bobby continued quietly, "don't ask any questions. Just fucking sign it. And smile."

He watched me silently until I faked a smile.

"And like it," he told me. Then he breezed away.

Bobby Lee was like that. When a roughneck rushed up to him hollering about a problem, he raised his hand slowly, palm down, to eye level.

"Do you see my hand?" Bobby asked.

"Uh, yeah."

"It's not shaking," he said.

That same day, a crane operator asked Bobby if he should get started on something.

"Did I tell you to start something?" Bobby spoke so quietly that he forced the crane operator to lean in.

"No," the operator replied.

"Do your job," said Bobby. He lit a cigarette, and then he walked off. The crane operator sat back in his seat and spit dip into an empty soda bottle.

Bobby Lee looked like Don Johnson. He wore blue jeans, cowboy boots, and a Western-cut blue denim shirt with a pack of USA Gold cigarettes in each breast pocket. He stood out, because on location everybody else looked the same. Faces hidden behind shaded safety glasses, heads covered by hard hats, the men wore fire-resistant jumpsuits with metallic stripes down the sides. They smoked, chewed dip, or spit sunflower seeds. They wore the same goatees, mustaches, or beards, dirt on their faces and dust in their teeth. They sweat and they cursed, and when they saw my green hard hat— I was a literal greenhorn—they identified me as a dumb fuck. But I couldn't tell any of them apart. Throughout the day, they appeared in bright sunlight in front of me, shouted a few words, and then disappeared back into the dust.

"I just started last week," a fellow greenhorn yelled at me over the din of a truck. "They don't tell you how to do anything. I don't know why. They'd save a lot of time if they told you."

"Don't stand there. Hand me that hook," a truck driver hollered at me. I handed him the hook, and he turned his back to me completely. I stood there for some time staring at his back. Then, unsure of myself, I walked off.

One guy told me to watch out for the crane operator I was working with. "He'll hurt you," he said solemnly.

"Can I throw this chain to you?" a dude asked.

"Yeah," I said.

"You ready?"

"Yeah," I said.

"You ain't ready," he said, and he threw the chain at me.

I thought of what the Safety Man had told me: "You're gonna be able to tell your grandkids you moved rigs across the Bakken Formation at the height of the North Dakota oil boom." I thought about how much money I was going to make.

On my first day, I had teamed up with another field hand, a young, muscular guy with braces. We rigged a twenty-foot-tall steel structure, called a derrick stand, to a gin truck. For the sake of simplicity, I'll just say right now that a gin truck is a big truck with a small crane on the back. It acts somewhat like a tow truck. We wrapped the chains from the gin truck around the steel structure and stepped away from it.

The driver pulled away, dragging the thing behind him. I followed.

"Don't stand under that derrick stand," the driver yelled at me through his window. "That's a dumb fucking obvious thing you shouldn't do. If it tipped you wouldn't get out from under it. How long you work here?"

"Today," I yelled back to him.

"Well, don't stand under the fucking stand."

I stepped away. The gin truck driver, a white-haired oldster with a handlebar mustache and a cranky voice, later approached me. "You started today? Well, I'll tell you when you do some dumb fucking shit you obviously shouldn't do."

"Please do," I said.

"I will. I'll let you know when you do dumb fucking obvious shit."

"Thank you," I said.

"YOU DON'T WANT to see that?" It was Bobby Lee again. Lunchtime, and I'd fallen asleep in the van. I blinked my eyes open, wiped a string of drool from my mouth, and looked out the window. The company hands were raising the derrick.

I climbed out of the van and stood beside it in the dust. The derrick is the tower that supports the network of pipe that is inserted into the earth for drilling. It gives a rig its height and makes a drilling rig look like a drilling rig. In the process of putting a rig together, the derrick is assembled horizontally, then hinged to the drilling floor, and raised on that hinge by braided steel cables spooled around a steel drum called the draw works. When erect, the derrick, like the mast of a ship (it is sometimes called a mast), sits atop the rig, directly above the drill hole, or well. The derrick on this particular rig, Sidewinder Canebrake 103, when completely raised, would stand over 160 feet in the air.

I watched the derrick pivot slowly upward. Behind it hung a vast swath of perfect blue sky. I took out my phone and recorded a video. I was bearing witness to the beginning of a process in which men would wrestle oil, one of the modern age's most precious materials, from shale deposits buried deep in the earth.

Oil was once thought to have mystical properties. In western Pennsylvania, the cradle of American oil exploration and production, Native American shamans from the Seneca tribe used the strange black substance in ritualistic healing exercises. They also used it to cauterize wounds and seal canoes. White men caught on, mixing what they called rock oil with liquor and selling the drink as a cure-all. The snake oil salesman was born.

In 1859, in the town of Titusville, Pennsylvania, near the home of the Seneca, a self-titled "Colonel" Edwin Drake drilled the world's first working oil well. Refined into kerosene, oil came to replace whale fat as the world's primary illuminant. Kerosene became "the light of the age." It literally made the day longer. And it made money. Oil was dubbed "black gold." A rush began. Oil City grew out of the muddy banks of Oil Creek, and the planet experienced its first oil boom.

It is strange to think of industry containing magic, but even in our current age of reason, it is hard not to think of oil as containing properties that, although also destructive, can only be described as utterly fantastic. Oil, refined into gasoline, diesel, and jet fuel, makes cars drive, planes fly, and ships sail. It powers submarines and spacecraft, sends shellfish to Colorado, medicine to Liberia, and soldiers and supplies to war zones around the world. It powers our school buses and gunships, ambulances and tanks. But it does more than transport us. It heats our homes. It illuminates our nights. We eat off it. Petroleum is contained in the glaze on our china, the finish on our tables, the linoleum on our counters, the tile on our floors. It wraps around us in the fabric that clothes our naked bodies, in the jewelry that adorns our necks and hands and fingers, in the balm that soothes our cracked lips, in the makeup that defines the contours of our faces. And it is inside of us. The syringes and the pills that deliver medicine into our systems are made from petroleum. Synthetic ammonia, derived from a process using refined oil, is an essential ingredient in fertilizer. Without it, farming on an industrial scale would not be possible. It is in our food. We eat it. Oil is so prevalent in every aspect of our existence that it has become all but invisible to the vast majority of people who need it simply to live.

And the people whose job it is to extract it from the earth? A group of them approached me. Dirty faces, goatees, and dark safety glasses.

"You ever been turtle fucked?" one of them asked. I braced myself.

"No," I said.

They shrugged and laughed and kept walking.

"Why don't you show him?" one asked.

I stood in the dirt and watched them move away from me. A rush of wind tore across the work site, peeling up a layer of dust and pushing it through the greasy machines, where it dissipated in reams of waving western wheat grass. My body was coiled tight, blood pounding through my brain. I took a deep breath, let the air fill my lungs and loosen my limbs. I spit on the ground and got back to work.

BOOK ONE

BOOM

—

Okemah was one of the singingest, square dancingest, drinkingest, yellingest, preachingest, walkingest, talkingest, laughingest, cryingest, shootingest, fist fightingest, bleedingest, gamblingest, gun, club and razor carryingest of our ranch towns and farm towns, because it blossomed out into one of our first Oil Boom Towns.

—WOODY GUTHRIE

RAIN

I'm driving across North Dakota in a red Chevy Blazer with my belongings piled in the back. It is spring. Rain pours from the sky, spills down the grassy buttes, and empties into the twists and cuts and curves in the land. The land is full of water, lakes called "kettles." The kettles churn in the wind, splash and slap against the macadam of the highway. The few trees by the side of the road are swallowed up to their ankles. The prairie can't hold all the rain, just like the town of Williston, North Dakota, can't hold all the men pouring in looking for work.

My Chevy threads its way south down Route 85. I see rows of RVs and unhitched house trailers set in muddy parks. I pass "man camps," columns of prefab housing that shelter out-of-town workers. On my right, my eyes linger on a junkyard swollen with wrecked cars and pickup trucks. I clutch the steering wheel with both hands and feel my own vehicle shiver from the gravitational pull of an 18-wheeler barreling past me down the narrow ribbon of highway. Tractor trailers rumble by, semis careening toward twilight. I drive past frack sites, drilling rigs, pump jacks, and flares. Everything I pass is born, lives, and dies all on the power of that

slick grease that now flows so sweet, so crude from the ground beneath those wet fields of prairie grass stretching into the horizon.

It is May of 2013, and I am joining what some say will be the biggest oil boom in American history. The road into Williston is full of
cars with out-of-state plates: Montana, Idaho, Washington, Tennessee, California, Texas, Florida, and more. The drivers look like me:
white men with tousled bedhead and lean, hungry visages, their
faces lined by sun and work—or else by the distinct lack of it. Everybody is new, kids on the first day of class, unsure where they are
going or how they are getting there. Through windows smeared by
spattering rain, I see their worldly possessions, likely all of them,
piled in the backs of dented station wagons, junker cars, and rusted-
out trucks.

At a stoplight, a man hops out of the truck in front of me and
runs back toward my SUV. Tall and thin and almost comically desperate, he asks for directions before I can get my window rolled
down.

"I just got here, too," I tell him through the glass.

WELCOME TO WILLISTON, ND, the sign reads. BOOMTOWN, USA.

PITHOLE

The boom spawned by the Drake Well in Pennsylvania in 1859 was modest compared to what happened six years later, fifteen miles south of Titusville, in a hilly area of near-unbroken wilderness where, as one story tells it, a spiritualist using a witch hazel stick "doodled the rod" and divined the location of a rich vein of liquid greenbacks, only feet below the ground's surface.

That well, known as the United States Well, would set off a bonanza comparable to the California Gold Rush. Within months, hundreds of derricks sat drilling within a few cluttered square miles. In 1865, the Civil War ended and thousands of men, including a surplus of veterans, flooded the valley. The area's population grew from fifty scattered souls to upward of twenty-thousand bodies within a single year. The wilderness was razed and, like a mushroom that grows overnight, a city was born. It was called Pithole.

Now North Dakota, like 1860s-era Pennsylvania, has boomed. Williston sits near the center of the Bakken Formation, a subsurface rock unit bigger than the state of California, and one of the largest contiguous deposits of oil and gas on the planet. The gourd-shaped Bakken Formation underlies the soil of North Dakota and Montana and stretches into the Canadian provinces of Saskatchewan and Manitoba. Advances in drilling technology—horizontal drilling and hydraulic fracturing—have turned this massive though previously unrecoverable shale deposit into a river of sweet crude grease. During horizontal drilling, a joint is inserted at the end of the drill bit, making it like a bendy straw. Hydraulic fracturing is an application wherein sand, along with a chemical stew called "salt water," is pumped through the drill head at pressures high enough to crack the strata surrounding a well. These technologies allow drillers to reach reservoirs that, until recently, would have been inaccessible. As of 2013, the United States Geological Survey (USGS) put the amount of recoverable oil from the Bakken at a mean estimate of 7.4 billion barrels. In layman's terms, that's a lot of fucking oil.

These technological advances along with the state's lax regulatory environment and a spike in worldwide oil prices combined to make investment in oil exploration in North Dakota incredibly attractive to wildcatters. They put the straw in the milkshake and started to suck. The rush to the region followed.

Oil field contractors began moving into North Dakota just as the working class of Middle America found itself unemployed. In 2008, the economy cratered, the housing market collapsed, and the price of oil peaked at $145 a barrel. With no houses to build, framers had nothing to frame, painters had nothing to paint, roofers had nothing to roof. The cost of silver and gold dropped, and miners had little to mine. The Great Recession flipped nationwide migration patterns on their head. While constricted economic circumstances led to the majority of Americans "battening down the hatches," and sticking close to home, the flood of workers into northwest North Dakota looked like a modern *Grapes of Wrath*.

Desperate for bodies to work the rigs, North Dakota's oil field companies gained a reputation for offering good pay, benefits, signing bonuses, per diems, and housing to any dude who could make the trek to town and swing a hammer when he got there. Men from all corners of the nation poured into the region looking for jobs. The population of Williston nearly tripled between 2008 and 2013, rising to roughly thirty thousand, with most estimates not accounting for job seekers sleeping in their cars or crashing in trailers and flophouses. Some estimates put the number closer to fifty thousand, others higher still. Between July 2012 and July 2013, one new person arrived in Williston every three hours—eight new people each day.

Who were these people? Men who *flooded* into Williston, *spilled* onto the streets, and *poured* in and out of bars, taking on the very attributes of the resource they had come to extract. Far from the delight of their children, far from the comfort of their girlfriends or wives, far from the comradery of brothers and fathers, or the caring of sisters and mothers, boots tottering on this ragged edge of America, so close to falling off the flat earth, with no money in their pocket and no friends for a thousand country miles.

Who were they and what were they running from? What was I running from?

I hit Williston with almost $3,000 in cash and about $2,500 in credit. You could say I was boomer rich, but in most ways, I was like everyone else. I didn't foresee a long future in the oil and gas business, but neither did plenty of other guys. Williston was a tool, and we were using it to extract money from the oil companies in the same way those companies were using us to extract oil from the earth.

Perhaps the most concise explanation can be found in an old joke I pulled from Smith Dalrymple's 1914 humor book *From Pithole to California*:

> Where is the best place to go when you are broke?
> Go to work.

All I knew was that I was driving into a city whose population was, as one oil field hand would put it, "ten pounds of shit in a five-pound bag."

It was a modern-day Pithole.

DK'S

DK's Lounge is a square, windowless brick building that, from the outside, could pass for a bomb shelter. It sits on an access road off the main strip surrounded in every direction by parking lots and squat one- and two-story buildings the color of dust. Inside is a cavernous, open-room casino dressed in dark velvet like a porn shoot from the '70s. The place is huge: the main room holds a circular bar, some pool tables, and a handful of blackjack tables. I brush past a guy on my way to order a drink.

"Excuse me," I say.

"Is it Feel-Me-Up Thursday already?" He is thin and craggy with a ton of pomade in his slicked-back hair. He watches me order a beer from a young bartender in heavy makeup and a tight, low-cut top. "Am I a pervert to stare at her cleavage?" he asks.

I shrug off the question but offer him a smile.

"My old lady is a pain in the ass," Pomade says. "She's hot for a forty-year-old, though. You could open a beer bottle on her ass." He actually whistles. "She's studying criminal law in South Dakota. Wants to become a cop. With my record!" he says. He leans back against the bar, swills his beer, and looks around the room. "Thing is, she's gonna be able to get shit on everybody. She'll wanna be a Goody Two-shoes for a year or so, but I'll make a dirty cop out of her. You bet. I'll turn her into a dirty cop."

He smiles, relishing the thought. "I spent six years in prison in Pittsburgh," he says.

I've always had great admiration for scoundrels and I've found, throughout my life, that they are drawn to me. I'm a good listener, and I'm not quick to judge. Until I've sussed out a situation, I keep my mouth shut. The boom, I am to realize, is like the first day of high school or college: everyone is new and hungry for comradery. I can use this to my advantage. I need a job.

A couple seats down from us, a guy in a Pirates ball cap invites himself into the conversation. The two of them talk baseball, and I listen. After another beer and some more bullshitting, I tell them that I am looking for work, and they give me my first inside education on boom economics, oil field jobs, and the kind of people who work them.

Pomade is an electrician. He works two weeks on and one week off. On weeks off, he drives to South Dakota to see his old lady and their nineteen-month-old daughter. He also has a son in Pennsylvania who is twenty-six. His son is married with kids, so Pomade has grandkids older than his baby daughter. "But that's life," he shrugs.

Ball Cap is a truck driver. Or about to become one. "Just started today," he says. He is pushing fifty, and he's been unemployed for some time, living off his wife's income in Wyoming. This will be his first trucking job, hauling water and sand to frack sites from 4:00 a.m. to 4:00 p.m. for fourteen days straight. He then gets a week off and returns to work the night shift. It will continue like that, with days and nights flipped upside down and a week off in between, for as long as he holds the job.

"I'm only making sixteen dollars an hour," he says, "but I get a guarantee that I'll be paid for one hundred and ten hours each week. And they promise I'll never work more than eighty hours a week. So it's a big win." This doesn't make any sense to me, but the companies in Williston are hard up for workers. I will learn that bizarre math is not uncommon in the oil field. "And," Ball Cap tells us, "I get paid for forty hours on my weeks off."

"What about overtime?" asks Pomade.

"Time and a half after forty," says Ball Cap.

"Double time?"

"With that many hours it doesn't even matter." Ball Cap takes a slug of beer. Pomade nods in appreciation.

A pair of women sit at the end of the bar. They are the only women in the joint except for the bartender. Costumed in dyed hair and heavy makeup, big earrings, and long fake fingernails, they sip colorful drinks and speak quietly to each other. Pomade gives them a sideways glance. "Thing is, you start making eyes with them and thinking they like you, but they're just sizing you up for money," he says.

Ball Cap takes his hat off and then puts it on backwards. He says his buddy was recently given two hookers as a gift for his birthday. "Twenty-six hundred dollars," Ball Cap says. "That's the kind of money that's up here. He had them all night, too. All night. From nine p.m. to nine a.m. They were out at the bar, hanging out together."

"I like hookers," says Pomade. "I've done a lot of drugs with hookers."

"All night," says Ball Cap. "His friend got him hookers for his birthday!"

Pomade nods in appreciation. "His friend is a good man."

I'm out of cash and trying to stick to a budget, but I don't want to leave. I stand for a while with nothing in my hand. Ball Cap notices and buys me a beer. I'm the only one without a job. "And besides," he says, "they're only three bucks for happy hour."

Pomade pulls a photo of his daughter out of his wallet. "It kills me to be away from her," he says.

"Oh, I know," says Ball Cap looking suddenly miserable. He shows us his phone's screensaver, a picture of two kids, ages five and six. "I just talked to them," he says. "They don't understand I'm gone. They think I'm hiding in the basement. They said, 'We can't find you.'" He looks at Pomade, then at me, then sheepishly back at his phone. "I play hide-and-seek with them," he says.

The women at the bar get up from their seats, scan the room for eye contact, and then move slowly through the lounge and toward the door. It was impossible to tell while they were seated, but standing it is obvious that one of them is a little person, barely four feet tall. We watch the two of them as they weave their way through the crowd of men. Pomade has his eyes on the small one. He squints at her and a low murmur of appreciation leaves his throat. "Mmm-mmmmmm," he groans.

A drunk at the bar stumbles up to us. He wears a black T-shirt and a nose ring; he has a shaved head, a thick beard, and gauges in his ears, and he wants to talk about drills. He wants to know what drills we think work best. He looks me in the eyes and wobbles. I don't have a favorite drill. Pomade is game, though. They debate the torque on DeWalts. Ball Cap wanders off to play blackjack.

The bar grows louder and drunker, and happy hour ends. I finish my beer and say goodbye to the men. I'll never see any of them again.

It is drizzling as I step outside. I turn up my collar and take a few steps into the parking lot when I'm startled by a group of drunks who explode out of the door behind me. "Goddammit, Chris, you better put your dick away!" shouts the loudest of the group in a high-pitched simulacrum of a woman. His friends double over themselves in laughter until another guy does his impression: "You guys are eighty-sixed!" Together they holler, in a shambling, liquid unison, "Put your dick away, and get the fuck out of here!" I walk across the street to my Chevy, parked behind a Super 8 Motel, and watch the drunks stagger to the next closest watering hole, the rear of the Vegas Motel's lounge. They pour themselves into it, and the door slaps shut behind them.

An old-timer stands on the sidewalk wearing a red trucker's cap and blue jeans pulled up to his tits. He is with a ruddy-faced kid, who barely looks twenty-one, and they are both in their cups. "Where's the entrance to the bar?" the old man asks. I point to the front door of DK's. He stares across the street and blinks several times trying to focus.

"Where you coming from?" I ask.

"I don't know a goddamn thing about anything around here," he barks, dismissing me with a wave of the hand. I watch him nearly fall over in the middle of the road as he and the kid stumble into the building on the other side.

I climb into the driver's seat of the Blazer. There is a six-pack sitting on the passenger side. I'd bought it earlier and left it there. I think for a moment, then twist open a beer and take a sip. My hand is trembling. Am I scared? I take stock. Yeah, I'm scared. But the beer tastes good even if it is a little warm. Whatever calms the jitters, cowboy. I take a big swallow and wipe my mouth. Then I start driving downtown. It is raining, and the beer washes my throat like the rain washes the streets.

FATHERS OF MEN

I arrive at J Dub's Bar & Grill, a sports bar with wood paneling and exposed brick walls set just west of the train station. In the bathroom, there are fist-sized holes in the drywall and steel plates bolted above the urinals. I slide up to the bar and order a cheeseburger and a beer. A big guy sits down next to me. He is young with an open red face. He says hello, then smiles and lifts his cap to reveal a sunburn that ends an inch below his hairline. Above it, his skin is alabaster white.

"So I gotta keep my hat on," he says, allowing a shy grin. He asks me how long I've been in town.

"Today," I tell him. "You?"

"I just walked here from the train station," he says.

We shake hands. His name is Buck, and he is from Missoula, Montana. He is the first friend I make in Williston. We will never become close, but for a short while our lives run parallel. We are both new to town. We need jobs, and we both face the ticking clocks of diminishing bank accounts.

Missoula Buck is anxious to start looking for work. It is Friday night, and he doesn't want to wait until Monday morning. "I struck out for work once to make something of myself, but my friends abandoned me," he says, "and I was homeless for a few months, living out of motels in Elko."

"Elko, Nevada?" I ask.

"There's a lot of work in the mines," Buck tells me. "Well, there was before they raised the mine tax. Now everybody's leaving."

"You were a miner?" I ask. He shrugs and tells me he did it for a while.

I tell him I am sleeping in my vehicle, and he gives me a concerned look. "It's packed to the gills where I'm staying," he says. "I dropped my bag off before coming here to eat. I've got my own bed, but a couple of brothers share a mattress on the floor at my feet. I'll let you know if it changes."

I nod and thank him. He seems genuine, even a little green. "I worked as a supervisor of the meat department in a grocery store," he tells me. "I figure I can always fall back on that. I checked in with a local spot, and it sounds like they might need somebody. Baggers there make fifteen dollars an hour."

A young man sits down to my left and politely orders a beer. He's got a ruddy, sunburned face, and when I nod hello, he holds me for a moment in a pair of steady blue eyes. "What's your work?" is the first thing he asks.

"I'm looking for work," I tell him.

"What are you trained to do?"

"Nothing," I say.

He works on the pipelines as a welder's assistant. "Drove out to the job site this morning," he tells me with the slightest trace of a smile, "but they didn't need me 'cause of the rain. I get twenty-six dollars an hour for the drive plus a per diem of a hundred bucks a day. So I got one hundred and fifty-two dollars today. Just for waking up."

"Sounds pretty good," I say, and Buck, next to me, agrees.

"Gravy," the welder's assistant says. There is a deliberateness to him, from the way he talks to the way he sips his beer. He says he'll be working from spring to fall in Williston, and then heading back to Coeur d'Alene, Idaho, where he works, like Buck once did, as a miner. "It's warm underground," he tells us. He talks about Coeur d'Alene, and in his eyes, the lakes shine and the forests shimmer.

He is living in a trailer north of town now, "next to my brother's trailer." That gets us talking about family. The welder's assistant laughs telling a story about his dad beating him with a belt. I grunt in acknowledgment.

"My dad whipped my ass, too," Buck says.

"I bet," says the welder. We all chuckle a bit.

"You ever been hit by a braided belt versus a solid leather belt?" Buck asks.

"Oh yeah," says the welder.

"There's a difference!" says Buck.

"Gotta say I prefer the braided." The welder smiles.

"Any day," affirms Buck. I listen and nod along.

"Once," the welder tells us, "my daddy had me pull a switch from a cherry tree. Boy, that was the most painful whipping."

"Them branches got snap," I say.

"I know all about switches," Buck says, "and all about soap. My dad'd make me wash my mouth out with soap. For cussing, you know. Once, he made me eat the soap, chew it, and swallow it. I got so sick. I tasted soap for days!" Buck laughs, and we laugh along.

This exchange proves to be the first in a pattern of conversations that I have with countless numbers of men in Williston, often upon first meeting. The conversation can be boiled down to two short sentences: "What kind of work you do? Man, my dad whipped my ass!" I come to think of it as The Williston Hello.

My father was a simmering, rageful presence in my life. Distrustful of his own wife and six children, he lived like a foreigner in his own home. It was always "your mother" and "you kids"; he spit out the words, refusing ownership of his own blood. To him, kin was enemy, and this paranoia made him ridiculous. He felt left out and put upon at the same time. Being raised by my dad was like being the ward of a mentally unstable teenager.

My father's father was a flophouse drunk, and while my dad rarely touched alcohol, he had all the attributes of a full-blown alcoholic. He was a leering, swaggering dry drunk—narcissistic, depressive, ignorant of personal boundaries, and prone to unpredictable mood swings. We lived in a ramshackle post–Civil War farmhouse built on a lush hillside on fifty acres of valley in central Maryland. The unruly beauty of our surroundings was staggering, but my old man treated the house with the respect an angry gorilla gives a cage. He pounded the walls with his fists, kicked against the plaster, slammed doors until they hung barely from their hinges. He ripped the rabbit ears off TV sets, broke the wobbly legs off chairs, and smashed plates and cups and dishes with such regularity that, as a kid, it was as if the whole world was cracked. My memories of these tantrums, like the dinner plates, are shattered.

One shard: My mother rushes the whole lot of us, my two brothers, three sisters, and me, out the front door of the house. My brother Ryan dropped his fork at dinner, and now the old man is howling behind us, the dinner table shaking from the force of his fists. We race across the overgrown green grass of the front yard and toward the station wagon, my mother urging us on like a track coach, "C'mon, c'mon, keep it up, let's go."

In these moments, we are completely isolated. Our driveway is a quarter-mile long, up a steep, jagged hill. Behind the house is a wide expanse of field, and then a forest. No neighbors in view.

My mother holds the car door and we pile into the back seat, the whole brood frantic—brother piled upon sister piled upon brother. As my mom slides into the front seat and turns the ignition, the old man appears framed by the doorjamb of the house. He was a boxer before he joined the army—he had a reputation as a man who could take a punch—and he moves toward us in that boxer's stance: shoulders hunched, fists forward, strangely light on his feet. His face looks gray, overcast, and somehow calm with an anger that belies expression. He steps onto the lawn and, as the car moves forward up the drive, he starts to run toward us.

I hear my younger sister's voice: "Oh no. He's coming, Mom, he's coming."

From the back seat I watch him, his face contorted now, howling a stream of inarticulate curses, spittle flying from his chin. He is maybe fifteen feet away when his boots hit the busted macadam of the driveway. He reaches down and picks up a rock the size of his fist. "Get down!" my mother yells. "Get down!" He brings back his arm, and he throws.

He misses. Through the window of the car, I watch him stoop again to the gravel. He picks up another rock.

A FEW YEARS after I leave Williston, I will find myself glued to the radio in my truck, running late for work but unable to stop listening to Pastor Joe Ehrmann, a former Baltimore Colts defensive lineman, talk about what he calls the father wound:

> I still have a father wound. A hole in my soul shaped like my father. A tremendous number of men in America and women, boys and girls, are wounded by absent dads or missing dads in their lives.
>
> The father wound is a big issue. It particularly plays out in masculinity because men are never really sure of their masculinity.
>
> Every man needs two things in life. One: they need some kind of father, some kind of man that's a little farther down the road that can look back and tell us that we're on the right path . . . and the second thing every man needs is some kind of brother that you can walk arm and arm as you struggle through life together.

That scar, that hole in a man's soul the shape of his father, was a defining feature of every man I met in Williston. Men had built their lives around it. Like a tree growing around a hatchet. The father wound served as a method of communication between me and the men I met. We talked jobs, then fathers. Before women, before politics, before home, "Man, my dad whipped my ass!" It bonded us together. Because we didn't have that man a little farther down the road, but maybe for a night, over a few beers, by showing off our scars, we could find some kind of brother.

Before I leave the bar, Missoula Buck nods his big sunburned head at me and clasps his meaty hand around mine. He gives me his number. "You can always use a friend," he says.

. . .

I HEAD TO Bill's Back 40. The bar is crammed with men in dirty work clothes. POW and MIA flags hang from the walls. A pop country song blares out of the satellite jukebox. It sounds like a macho deodorant commercial. I love the outlaw guys—Waylon, Hank Jr., and Billy Joe Shaver—but in 2013 bro country is at its peak. I push my way through the crowd. At five foot seven and 150 pounds, I have never been big, but I have never felt as small as I feel crossing that room. The men crowding the floor—and there are only men, or boys really—stand heads higher than me, wearing PBR- and UFC-branded caps, stained Carhartts, and dirt-smeared jumpsuits. "I'm the Red-blooded, Hardworking, God-fearing American Obama Warned You About," one shirt reads.

I meet an Australian in a braided cowboy hat with the brim curled over. He has thick sideburns. "I won the Wolverine contest!" he hollers at me over the music.

We end up blind drunk in his souped-up pickup truck, bouncing and rumbling through the night. The evening is wet and warm. The air smells fresh and sweet like a melon just cut open. The Australian tells me that he slept in his truck when he first got to town. "No shame in that," he says, driving me back to mine. "This is history, mate!" he shouts out the window, to the road and the prairie, the oil and the money. "The fuckin' gold rush!"

BOOMTOWN, USA

I have just enough room to lie flat from head to toe. I can't roll over—surrounded as I am by camping gear, clothes, my guitar—a parlor size Blueridge acoustic nestled safely in its case—books, and clutter. Light filters through the tinted windows of my Chevy Blazer. Morning. My mouth is dry, and I need to piss. I blink myself awake to the sharp voices of children. A mother and father walk past, each holding the small hand of a child. I watch them from behind the dark panes of my vehicle's windows.

I piss into an empty Gatorade bottle, set it by my feet, and go back to sleep. When I wake up again, wrestling a hard-wired sense of shame at being seen clambering out of the rear of my SUV, I flip open the back window and unhitch the tailgate. When it drops, I slide out on my ass. Boots hit asphalt. I stretch in the hot North Dakota sun.

WILLISTON IS ROUGHLY four miles across at its widest point—from the Walmart to the Amtrak station. It looks like any ailing midwestern town . . . during the Carter administration. Fast-food restaurants, two-star motels, car dealerships, and box stores, depressing just to list, line the approach as I pilot my SUV down Route 2's "Million Dollar Way."

The city's downtown, with a movie theater proclaiming "Our Screen Talks," has a certain rustic charm. It is all but deserted . . . wide, flat streets, a coffee shop, a bus stop, a bookstore, and a billboard with the Ten Commandments displayed just outside the two topless bars at the foot of town—the perfect place for an insecure god to rage against greed and sex and killing. A traveling stripper and journalist, Susan Elizabeth Shepard, would later write of Williston that it was less than a town, "something closer to a military base or the world's saddest campground." She was right. I drive in circles. These first days the roads tangle in my mind like a plate of spaghetti.

I GRAB A SEAT at the counter in Gramma Sharon's Family Restaurant. I make small talk with a plump waitress, eat chicken fried steak smothered in gravy and covered by two over-easy eggs, and drink cup after cup of hot, flat coffee. I text Missoula Buck. He tells me of a company that is hiring and gives me the name of the receptionist.

I wish him luck on his own job search. Then I drive to the library. Cars and trucks stuffed with clothes and bedding sit in front of signs that read NO OVERNIGHT PARKING. I walk past a man snoring openmouthed in the back of a hatchback before heading inside.

Big-boned white men sit alone at scattered tables like lonesome rhinoceroses, poring over job applications and hunting-and-pecking online forms. I loiter, researching job leads on the internet for as long as I can stand it. On my way out, I stop in front of a large display by the exit. It features rows of photographs—mug shots of dour, slack-jawed men staring at the camera: local sex offenders, their names printed beneath their pictures. A few years of felon-friendly, no-questions-asked hiring by oil companies has gifted Williston, North Dakota, with the highest concentration of rapists and child molesters in the nation.

I wouldn't know the numbers until later, but the crime rate in Williston was four times the nation's average. Since the start of the boom, assaults had skyrocketed times twelve while thefts had quintupled. Reports of rape had increased sixfold.

The playground outside the library is built to look like an oil lease with a teeter-totter shaped like a pump jack. The small park on which the playground sits was, not long before, the site of a tent city, a migrant encampment before the city evicted the squatters. Now it is empty. I sit on a bench. There is something that I can't quite put my finger on—an unsettled feeling. My hackles are raised. I don't know exactly what it is, but I feel it in the hollow of my stomach—*something* is "off."

After I left the family farm, I spent ten years in Baltimore City. By the time I got to Williston, I'd had a gun drawn on me twice, once by a teenager and once by a kid not old enough to be called a teenager. I'd been in other thorny situations as well, and the unsettled notion I experience in Williston is something I can only compare to the anxious feeling I got taking a wrong turn off South Bentalou Street in Pigtown at dusk, or standing alone at a bus stop with a broken light, drunk one night south of North Avenue, watching two figures approach me in the dark. In Baltimore, a place more famous for its crime than its crabs, this feeling was isolated to pockets, limited to certain locations, situations, and times of day; in Williston, if my stomach is right—and I trust my gut—it is *everywhere*.

A drifter shits on the floor at the rec center. A knife fight breaks out in a man camp. A bouncer at a local club discharges his firearm to break up a fistfight. A man is gunned down and killed on the street outside a strip club called Heartbreakers. An eighty-four-year-old woman is raped in her home. A twenty-two-year-old man is raped in his truck. Prostitution rings, Mexican cartels, methamphetamine and heroin traffickers have descended upon northwest North Dakota. Like me they are following the money.

"I wouldn't say it's out of control," Williston's mayor tells a reporter from CNBC. "But we're very close to that."

Sleeping in my Chevy—in these first unknowable days—I'm particularly exposed, and every barfly and truck stop cashier I meet has an opinion as to where I should park. Not on Main Street: "You'll get your windows busted out by the drunks." Not too far off Main Street: "A lot of creepers around."

I hear the same advice everywhere I go: "Be careful out there." They say it like it doesn't matter how careful I am. It gets weird fast: "Cartels have taken over the entire second floor of the Vegas Hotel. They're trafficking whores," a man tells me at Applebee's. "Three men are ass raped every day in this fuckin' town, I swear to God. I'm ex-police, I know!" It is strange as a man to hear that I should never leave my drink unattended at a bar, but I will hear it constantly in Williston. Someone is going to put something in it. They will. Don't leave it here. Take it to the bathroom with you. Seriously. Stories of man-on-man rape permeate the honky-tonks and dives—everybody has heard from somebody that they knew a guy who found a guy waking up in a parking lot with his pants around his ankles, bleeding from his asshole. Years later, I'll relay this experience to a girlfriend. "Now you know what it's like to be a woman," she'll tell me, her eyes flashing angrily. I'll nearly fall out of my chair.

In Williston, it is hard to separate the facts from the fiction. But the hyperbole has a way of creeping under your skin worse than the reporting does, and even without the exaggerations, all you have to do is take a look around. Testosterone-fueled young men are working fourteen-hour shifts at jobs that can kill them in a town without friends or family. It only makes sense that the dangerous work of the day spills past sunset as parentless white boys roar through the night in F-250s with no girls to fuck, but plenty of guys to fuck up.

"Just watch your back," a bartender warns me. "And keep your truck locked."

Night falls slowly in North Dakota. The sun lingers on the horizon until 9:30 p.m. in early June, and darkness doesn't come until well past ten. I look for a place to park the Blazer so I can sleep, cruising the suburban-like web of streets off Main.

I watch the lights in the houses flick on and then off as night grows ever longer. When it is very dark, I step out of my vehicle and shut the driver's door very quietly. I walk around to the back, drop the tailgate, crawl inside, and pull off my boots. I think of the families in the houses—husbands, wives, and children: they have jobs to work, classes to attend, property they've obtained, and lives to protect. I got none of that. I'm unloosed. Free, I guess. Alone, for certain. And my bladder is full. I piss into an empty bottle, but I have to stop myself before my urine fills past the brim. Tightening my muscles, I struggle to hold in my piss and not spill the bottle on myself as I screw the top back on and rummage around my camping gear for a second bottle. I fill that one up, too. With two bottles of hot piss at my feet, I lie on my back and close my eyes.

I've been crashing in my vehicle for over a week now. *Fuck this*, I think, *I need a real place to sleep*.

THE FLOP

I t takes a moment for my eyes to adjust to the lack of light. The
front room of the townhouse is dirty, and the shades are drawn.
Rumpled sheets and blankets cover the couch, and an unmade mattress lies on the floor next to the front door. A third sleeping space
has been thrown together against the opposite wall. Chewed dog
toys and frayed, dirty rugs cover the carpeted floor. A large-screen
TV blasts light and sound. People on a reality show stare out from
the screen, bitching.

"Here's the place," says Champ. A beer fits in his hand as if his
hand had been built around it. He takes a swallow. "New York!
Fucking New York!" he bellows. "Where in New York you from?
Brooklyn! Fucking Brooklyn. That wins you big points. I'm from
Jersey."

A bald guy wearing a work uniform and a squiggle as a frown sits
on the couch watching TV. "This is Frank," Champ says. "He's not
happy with me, bringing another guy in."

Frank looks at me and blinks. His face is smooth like a baby's and
his eyebrows are such a light blond color that they disappear against
his pale brow. I blink back at him.

"'Sup?" says Frank.

Champ plants his feet in the middle of the townhouse living room. He's in his midthirties, but the way he commands space makes him feel older. He has the build of a bulldog, square with broad shoulders, thick arms, and a big belly. He cocks his head to the side and addresses the bald guy. "Hey, I know it's tight. We just gotta do what we gotta do."

He turns back to me and shrugs. "It's better than sleeping in your fucking car, bro."

Champ moves through the living room. His bright orange hair is spiky, cut short enough to expose the sunburn on his scalp. He wears no shirt, and his face, neck, and left arm are painted red in a classic trucker's tan. His shorts hang low under his belly, exposing the tops of his hips. We walk past a toilet in a laundry room and into the small kitchen. Junk mail, magazines, and clothes cover the dirty table.

"Frank is from the Philippines or some shit," he says. "The other guy works third shift. We call him Jesus." Champ pronounces the hard *J* and continues, "He's a black guy from Africa. Upstairs I got a couple Jamaicans and a mother and her son."

We step out the back door into a small yard with a picnic table, a gas grill, and a hot tub. Champ stops. He spins around and squares up.

"No fights," he says, making direct eye contact. "It's okay to drink, have a couple beers, but no fights. We had an altercation a few nights ago between Frank and Jesus, but mostly everybody gets along. It's cramped, guys are tired—they had an altercation. I was out of town, so Rickie had to deal with it. This is Rickie. She's my girlfriend." A woman in her midtwenties with a pretty, round face framed by black bangs sits at the picnic table, sipping a Budweiser. She gives me a smile.

Champ continues, "I had to go home to Jersey for a funeral, and they had an altercation. I came back. I said, 'Do you fucking guys wanna leave, because I'll throw you out right the fuck now?' They both said, 'No.'" He shrugs. "There's nowhere to stay here, especially this cheap. It's not ideal, but it is what it is."

He looks directly in my eyes and points at my chest. "No smoking meth," he says. "No cooking meth. You can roll a joint, but don't smoke it in the house. It's okay to drink, but don't start any fights. And no racial shit. Okay? We all got to get along here. The Jamaican guys upstairs, they're really working out. They keep to themselves. No problem with them. I can never remember their names. This guy and his mom, they got Parkinson's or something. So don't be freaked out. They're all twitchy."

"They've got Tourette's or something," Rickie says.

Champ says, "They're real nice people." Then, sideways, "Church people, you know?" He shrugs. "We've got a lot of people here. Everybody does their best."

He walks me upstairs and shows me a small, stained bathroom— toiletries all around the sink, towels on the floor, mildew in the tub. There are three upstairs bedrooms in the townhouse. Champ and Rickie stay in the biggest room, facing the street, the mother and her son share the second bedroom, and the Jamaicans crash in the back.

I follow Champ back out to the patio and take a seat at the table with him and Rickie. Rickie asks me if I want beer or water. I take her up on the water. I know I look ragged. I can see it reflected in her eyes—the slight look of concern, the trace of maternal kindness. At first Champ and Rickie were sussing me out for danger, but they can see I'm not that. We talk about New York. "You get big points for that," Champ says, his broad, sunburned face turned toward me, "for being from the East Coast."

"I haven't seen too many of us out here," I tell him.

"We're a long way from home, buddy." I am 1,826 miles from home. With the exception of a cousin in Iowa, the closest person I know lives in Boulder, Colorado. One friend within a thousand miles. My experience of this circumstance vacillates wildly between an exhilarated sense of freedom, a true American road lust, and a fear so stark I feel it coil around my bones and watch it age me in every mirror I pass. This is as alone as I have ever been in my life.

Champ continues, "Rent is four fifty. It's midmonth. So if you wanna move in tonight, it'll be three hundred. No lease, no deposit, we'll set you up in the living room."

The oil boom has driven up the price of everything; but by far, the most expensive thing in town is housing. There are simply not enough places to stay. Rent in northwest North Dakota in June of 2013 is higher than rent in Manhattan. It will soon surpass San Francisco to become the costliest in the nation. One-bedroom apartments cost over $2,000 monthly. Single rooms in shared apartments cost over $1,000 a month. Some postings advertise women-only rooms for as low as $500 a month with sex as an implicit part of the deal, which is obviously out of the question for me. I had been scanning ads every day, and I'd found exactly one space I could afford.

"I don't see that I have too many options," I tell Champ.

"It's better than sleeping in your vehicle," he tells me again.

I have been sleeping in my Chevy for ten days. I thought I'd get used to it, but instead it was costing me money, wearing down my health and my nerves. I was eating out for every meal, fatty junk food mostly, and with nowhere to go at night, I was drinking too much, spending too much money on booze in an attempt to calm the nerves and sleep. I'd been washing up at truck stops, and I was growing more disheveled each day, a little less washed, a little more walleyed. I had just begun searching for work, and I knew if I was viewed the way I was starting to feel, it would be game over. Like desperate men don't get laid, poor men don't get hired.

I tell Champ I will take the place. I give him $300 in cash, and he handwrites me a receipt. "What's your last name?" he asks.

I had introduced myself as Mike. I tell him my last name is "Smith" and he stops writing. He twists his neck to the side and I see the trust race from his eyes. "Are you on the run? You're not a fucking fugitive, are you? We had one of them once. I don't want any—"

I laugh. "If you want," I tell him, "I can show you my ID."

"No"—he puts his hand up and ends the conversation—"I don't."

He finishes writing out the receipt and hands it to me. "Well, you're one of us now."

A PICNIC

Frank bustles through the front door of the flophouse carrying grocery bags. He wears a blue polyester work uniform and cap, both embroidered with the Pepsi logo. The cola company has a plant in town and Frank works there. A feral-looking kid, maybe twenty years old with a muskrat mustache and a shaved head grown a little too long, trails after him. The kid is somewhat tall and very thin with a dirty, oversized T-shirt draped over his string bean frame. He has a bandage on his hand and wears large wraparound sunglasses.

"Whassup, B?" Frank says to me. "This is my friend." He nods toward the kid with a hint of pride. "He started Pepsi yesterday."

The kid nods silently. I nod back, and Frank invites me to eat with them. They head out to the grill, and I grab a few beers from the fridge. On the patio, I hand them beers as Frank smears cheap barbecue sauce on fat chicken breasts. He slaps the chicken on the grill, and it sizzles in the sun.

Skip and his mom, Tammy, step outside. Skip is good-looking, in his early twenties, with styled hair, a goatee, and a bright disposition. He has Tourette's syndrome, which causes him to shake constantly, his arms sporadically flailing at his sides. Sometimes his neck pushes forward and his shoulders roll, but he also carries himself with a certain dignity. Skip's mom Tammy also has Tourette's, but her shaking is tighter than her son's—she keeps it closer to her body by swaying back and forth constantly.

"It's a blessing," Skip says. "Champ really takes care of us. Fridges in the rooms, two to a room, TV. It's not bad. I was living in Minot in a trailer in winter. No heat, no running water, and I had to take my mom in. She had surgery, and I had to take care of her. So, we're in a trailer, beds right together. God makes trials for you, but you get through the trials and things get better."

The night before, Skip and I had watched *Frasier* on TV. Skip laughed so earnestly at the corny jokes that I found myself laughing along. "He's so uptight!" Skip said about Frasier. "What shows do you watch?" he asked.

FRANK TELLS SKIP AND TAMMY that he has enough chicken to go around, and invites them to join us. They have a box of mac and cheese and offer to make coleslaw.

I take a seat at the patio table next to the kid with the sunglasses and the bandaged hand. There is something about him I immediately like—a sly intelligence. He puts a generous dip in his lip, spitting occasionally into an empty Pepsi bottle, and we get to talking.

"What happened to your hand?" I ask, pulling on my beer.

"I was working at B&G when a pipe came off the line," he drawls, fiddling with his bandages. "I picked it up too fast with my left hand and burned these fingers." He opens the palm of his non-bandaged hand and displays two nasty burns crossing the joints of his index and middle fingers. "Then I went to toss it away, and it stuck to my right hand, took the skin right off it."

"You go to the hospital?"

"Not right away."

"Did B&G pay for the visit?"

"They fired me," he says, spitting into his Pepsi bottle. "They said it was my fault because I wasn't wearing gloves. I bandaged it myself, but it blistered up real bad. I had a blister the size of a golf ball on my hand, two golf balls! I wasn't gonna go to the hospital, so I popped the blister on a tree." He smiles and mimes whacking his hand against the tree. "After that I said, I'm going to the fucking hospital."

Once a week, he went to the hospital, and they scrubbed his hand with a brush. "They gave me morphine, but it still hurt like a bitch," he says. "I'd go in there and they'd scrub it, get like half a cup of blood every time they did it."

He plays philosophically with his bandage. "They say it doesn't need to be scrubbed anymore." He scans the table, says, "I'm hoping that opinion holds." When he smiles wide, I can see the chew in his bottom lip.

I offer Skip's mom a beer and realize she has a pronounced stutter when she responds with a kind of joke. "I don't drink beer," she says. "If you drink beer that makes you an alcoholic. If you drink wine then you are a wino. That's why I drink hard liquor. Because I don't want to end up in AA." It takes Tammy a long time to get the words out. Her voice quivers with her body. When she finishes, she repeats the joke, then she repeats it a third time.

"My mom's an alcoholic," says the kid with the bandaged hand, "and she's been going to AA for twenty years. She last got arrested driving with a bottle of Bacardi between her legs, so I'm pretty sure alcoholics drink hard liquor."

The chicken comes off the grill, and big bowls of mac and cheese and coleslaw are brought outside to the table. I grab more beers, and we pass around the food until all our plates are brimming with it.

Frank, the kid, and I dig into our food, while Skip and his mother pause to say a prayer. I stop eating in deference to the faithful, but Frank and the kid don't even look up. They eat like the chicken might run away, shoveling mouthfuls into their open maws, their noses inches from their plates. Skip and his mom sit with their eyes closed and their heads bowed. The stutter and sway of their Tourette's give the prayer a strange, incantatory power—an electricity that raises the hair on my arms. When they finish, I whisper Amen—I was raised Catholic, I can't help myself—and, under the sweet, flat gaze of a springtime North Dakota sun, we eat. The coleslaw is cold and spicy, the mac and cheese gooey, and the chicken falls right off the bone.

Skip starts telling everyone about growing up in Jefferson Parish, Louisiana. His mom had a house in the ghetto, he says, and drug addicts would knock on their door at all hours, day or night. "It was scary," he says. When Hurricane Katrina hit the coastline, he fled with his father's family to a hotel in Houston, and Tammy ended up in a FEMA trailer in Mississippi. "But there was something wrong with the trailers," Skip tells us, "and people were getting sick. Mom kept bleeding, couldn't stop bleeding."

When they heard about the boom in Williston, they moved together to Minot, where they worked as clerks at a grocery store, but a manager fired Tammy for taking expired food home from work.

"Everybody takes the expired food home," she stutters. "Everybody gets to, but she says to me that I gotta throw the bad food away. Just 'cause she didn't like me. It's personal. It's all personal. Others got to take the bad food home."

"She's a cunt," Skip says. "She's just a cunt. She's mean to everybody. She's a cunt." He repeats it over and over. "She's a cunt," he says. "She's just a cunt."

"Just 'cause she didn't like me," his mom says, shaking in her seat. "It's personal."

We finish our meals, wipe our mouths, and lick our fingers for a moment in the fat silence of a shared full belly.

"This place is a blessing," Skip says.

HOMES

The kid's name is Jesse. That evening, he sits in the middle of the living room on a new inflated air mattress, eating a second dinner of a fast-food cheeseburger and fries, drinking a large soda, and silently watching TV. Scrawny with his shoulders stooped, he is the very picture of contented poverty.

He had no money for rent, but he asked Champ for a place to stay anyway. "No fucking way," Champ told him. But Champ sighed when he later told me the story, "Then Rickie comes to me with tears all in her eyes, 'He's just a boy,' she says. 'Maybe you can make an exception, blah-blah-blah. He's a baby . . .' So, I told him he could stay as long as he paid up by the end of the month."

It was a decision Champ would regret. Jesse, although he had no other options at the time—he'd been sleeping on the warehouse floor at Pepsi—would regret it also. And so would Rickie. But at that moment, four of us share the living room—Frank, Jesus (whom I have yet to meet), Jesse, and me. Ten of us share the townhouse.

A story about Sandy Hook plays on the television. It had been six months since twenty-year-old Adam Lanza entered Sandy Hook

Elementary School in Newtown, Connecticut, and shot and killed twenty children and six adults in the deadliest mass shooting at a grade school in US history. On the TV, a pundit makes shrill declarations about the dangers of gun control legislation. On the couch, Skip says he thinks the killings were orchestrated by President Obama. Frank agrees. I've never heard anything so weird.

"That's crazy," I say. "You think the president of the United States set up a guy to kill a bunch of schoolchildren so he could push through legislation to 'take your guns away'?"

"Think about it," Skip says. "It's the perfect plan."

"But it didn't work," I say. "He hasn't taken anybody's guns away."

"Hitler took away everybody's guns and that's what Obama wants to do," Skip tells me. Then he says, "Think about it. Hitler tickled the ear of the Jew. Now Obama is tickling the ear of the Jew."

I have no idea what that means, and I don't pursue it.

Later, Skip starts talking to Frank about the women in Minot. "They're married, they're fucking black dudes, or they're lesbians!" he says. Then he says it again louder and prouder, "They're married, they're fucking black dudes, or they're lesbians!"

Skip eventually goes to bed, and Frank stretches out on the couch, still wearing his Pepsi uniform. I lie down on my air mattress on the living room floor.

I grew up in a big Irish Catholic family. My parents had seven children. Brigid, the first born, died when she was very young. After her came Shanon, Ryan, Matthew, Megan, me, and Kate. We are all roughly two years apart. My mom's best friend was the matriarch of a similarly sprawling Irish American family, and as a kid I thought everyone on earth was Catholic. We lived near a town called Damascus. For years, I believed it was the location where the scales fell from Saul's eyes and he became Paul the Apostle. That story has always had a profound effect on me.

One night, when I was six years old, my father attacked my brother Ryan, who was twelve, for climbing into bed with a dirty shirt. "He told me I should change it before bed," Ryan later told me, "and I told him I wanted to wear it. Dad then said he was going to take it off me. There was a struggle, and I wouldn't let him have it. Then he said he was going to beat it off me. He ripped it off my back." My mother rushed into Ryan's bedroom—later she would say she thought Dad was going to kill him—and they made a run for it. My father chased after them, throwing books and shoes and anything he could get his hands on.

I'd grown used to the sounds of calamity—my father yelling and punching, my brother crying, my mom pleading—through the thin walls of the room in which I slept, but this night was different. The fighting drew my eldest sister, Shanon, from her bedroom, and along with Ryan and Mom, they fled the house in their underwear, racing across the yard and tumbling into the car. This time, Mom didn't stop driving when they reached the top of the hill. She took them all to a shelter for victims of domestic abuse.

No one seems to agree how long it was before Shanon's boyfriend Donny picked up Matthew, Megan, Kate, and me. Megan thinks it was weeks later, but I feel like it was only a couple days. None of us recall what we ate, but everyone finds it hard to imagine Dad cooking for us. One thing we all do remember is that we pulled the mattresses from our bedrooms into an unused area upstairs and spent endless unsupervised hours racing through the hallway, diving, jumping, and flipping onto the pile of mattresses. Van Halen's "Jump" had only just been released, and the DJs on DC101 must have spun it every fifth song.

We spent the rest of that summer on the run. There weren't enough beds in the shelter we first moved into—everyone shared a single room except Kate and me. The two of us slept in a pair of large cradles set up in the living room. I received my first kiss there from a young girl, maybe eleven or twelve, whose family was leaving. I woke in the first hours of morning to feel a full mouth wet on my mouth, to blink open my eyes and to see her slowly pulling back from me, her long hair tracing my cheeks, a plaintive expression on her face, thunderstorms racing across her eyes. I'd never see her again, but the memory would haunt me into adulthood.

Later we hid out at a farmhouse owned by a family friend of a friend. At the farmhouse, I had dreams about witches. I spent a terrifying day lost in a cornfield. One day, Kate and I came across a downed small-engine plane. There was no sign of a pilot or a passenger—just an empty plane in the middle of a cornfield. Kate set up a lemonade stand at the juncture of the closest road, and we practiced running away from the stand in the event that my father appeared.

My dad had always kept a gun in the house, a snub-nosed .38 revolver. Usually, it stayed hidden, but he made much use of our knowledge of it. "You just try to take me off my land!" he'd growl. "They'll be dragging dead bodies off this motherfucker!" That summer, the thought of the old man appearing with the gun and killing my mother, my siblings, and me was never far from any of our minds. For me, although it would waver in probability, it would remain a possibility up until the day he died in December 2019.

At the end of that summer, Mom moved all of us into a small rancher in a suburb maybe twenty miles from my dad, and she got a job making sandwiches at a local deli. Sometimes we'd rent a VCR for the weekend and watch movies based on Stephen King novels like *Firestarter* and *The Dead Zone*. These were some of the most fun years of my childhood. I had friends who lived next door. I learned how to ride a bike there. It was a good home.

At first, I didn't see my father often, but it didn't take long for things to fall back into a sort of rhythm. Abuse requires constant short-term memory loss. Dad had threatened to kill us, yes, but it wasn't the first time; so we did what we did with his previous threats, which is to say we didn't talk about them. We barely talked about the time he attempted to break into the house and kidnap my sister Kate and me—Ryan and my mom barricading the door to the basement as he smashed a window and tried to force his way inside. For years, I'd question if I'd ever heard him say the things he said, do some of the things he did. In hushed conversations with my siblings, sometimes after a few drinks or a smoked joint, one of us would broach the subject, "Do you remember . . ." And the flood of relief I'd feel would be near instantaneous. "I *do* remember" being three of the world's most powerful words. But time would pass and that learned forgetting would take over again, and again I'd question my memory until the next hushed conversation, sometimes years later. That cycle started early.

Not long after moving to the suburbs, I began spending some weekends at the farm, tromping through the woods, splashing through the creeks, and scrambling up the ageless trees. For the most part, Dad stayed on his best behavior. He often bribed Kate and me with presents and candy. And for me, it was the best of both worlds, because I adored the freedom and solitude of our untamed farmland, and I also got the chance to enjoy suburban life, playing kick the can with neighborhood kids.

They were tougher times for my brothers. Ryan was a sensitive kid. Before we left the farm, a classmate had called him a "faggot."

When he told my parents about it, my father responded, "Well, are you?" Ryan didn't know what the word meant. The church we grew up in sure did. He'd been shuttled between various psychiatrists and psychologists since he could remember, and he'd become prone to angry outbursts himself. He could be physical. More than once, Ryan attempted to yank the steering wheel while my mother or father was driving. I was in the back seat one of those times, and he nearly crashed the car. At our house in the suburbs, he got into it with Mom; she called the police and Ryan was locked up in a detention center for two weeks. Mom spent every day on the phone trying to get him out. He was released, but then he was committed to a mental hospital for a psych evaluation. He was fourteen. "I was surrounded by people with real, super-fucked-up problems," he'd later tell me. "After thirty days, they basically said I was fine and let me out." Emotionally, he withdrew. Matthew, naturally more voluble, exploded. He'd starting using drugs before he was in his teens and fell in with a group of neighborhood degenerates. Matthew's buddies stole cars, smoked cocaine, and burglarized houses. He was eleven! By the time he was thirteen, Matthew was in reformatory school. Both my brothers ended up moving back in with my dad.

After a few short years in the suburbs, in fact, my mother moved all of us back. The farmhouse that was my home for most of my young life had been built in the 1880s. In a constant state of disrepair, it was set down in a valley bursting with honeysuckle and poison ivy, brimming with groves of pine, spruce, and oak. When it turned sticky with humidity in the summer, the goats would climb over the broken fences. They'd clamber onto the front porch to chew the ivy off the porch beams and then charge through the front door when my mom ran out to scare them off. She'd race after them through the downstairs living room and kitchen, waving a broom and shouting as they bleated and shit all over the floors. In the spring and fall it was Eden with the serpent. In winter, we huddled around space heaters. I'd wake up to see my breath crystallizing in front of my mouth, shivering sometimes from the cold. The windows were like tissue paper, and the pipes froze every year. It wasn't unusual for us to lose running water.

When I was eighteen, I left. I moved to Baltimore and rented rooms in group homes in the city. They were the only places I could afford, filled with strangers living communally. I had learned to live quietly, not to assert my personality too strongly into any given situation. This made me an easy person to get along with. I remember sharing one house with a black cross-dresser who wrote Gothic folk songs. He frightened me, frankly, but I was used to being frightened. We'd sit in my room sometimes and play guitar for each other. A mother and son lived in that same house. The mother had recently left a mental institution. The son was a custodian who worked nights. This was in the mid-90s, and he showed me photos of a beautiful Russian woman he had met through the back pages of *Rolling Stone* magazine. He planned to marry her. I never let my judgment show. I lived among them like a friendly ghost. Much like when I landed in Williston, I stayed wary, not giving anything away to anyone until I was completely certain of where I stood.

IN WILLISTON, I take a good look around the room of the flophouse. I'm lying on a mattress in the corner, a bag of luggage at my feet. In the fading light, I see my new companions, spread over the couch and draped across the floor. They slowly drift toward sleep. I feel the spark of a smile grace my lips. The ridiculousness of it all. In a strange way, this place feels like home.

THE DREAM

I would come to think of Champ's place as a microcosm of the boom. The flophouse was a revolving door for people who came into town looking for a new start, a new job, and a better life. We had little in common, and yet we were living under one roof in one town at one specific moment in time because of vast historical and technological forces well beyond our purview; but, if you asked any of us why we were there, the answer would have been uniformly the same: we were there to make money.

"Ours is a speculating race," mused a writer from the *New York Journal of Commerce* in 1865, speaking of the population of Pithole, "prone to the cultivation of that art which gets sunbeams from cucumbers . . ."

Colonel Edwin Drake, the godfather of oil exploration, earth's first boomer, the first man to ever raise a derrick, was a dreamer. Considered a lunatic by locals, he was heckled while attempting to drill the world's first oil well. Drake kept at it. When he ran out of money, the company that had hired him offered nothing more than a train ticket home.

"You all feel different than I do," he wrote them. "You all have your legitimate business which has not been interrupted by the operation, which I staked everything I had upon . . . I got skin in the game," said Drake, "and I got a dream." He went on to irrevocably alter the course of human history.

And what of us? Piled onto mattresses in a flophouse in Williston, North Dakota? What sunbeams do we hope to extract from cucumbers?

Skip wants to get a job at a gas station so he can buy a reliable car. Tammy wants a shift at McDonald's. Frank plans to stick it out at the Pepsi warehouse so they will fund his training for a commercial driver's license, so he can make a middle-class living driving trucks. Champ hopes to pay off his house so he can sell it before the boom goes bust and buy another house somewhere nicer. For me, at this point—before I learn what it means to be *a good hand*, before I understand how important that concept is to me, how much magic those words conjure in and unto themselves—I just want to make some money so I can start over. In retrospect, it feels ridiculous, but I figured I'd work for three months, make $20,000, and get the fuck outta town.

My thoughts on what I'd do with that money are almost too embarrassing to record. They were the pie-in-the-sky ideas of a complete idiot. Embarrassment crawls up my neck just thinking about it. I was going to use the money I earned to fund development of a movie I was writing (I would never finish the first draft of the script, by the way), or . . . I was going to move to Thailand to study a meditation technique called qigong. Truthfully, I hadn't thought ahead. I just thought, no, I knew, that *money* would solve my problems.

Money was always tight growing up. My father had trouble keeping up with the taxes on our house, and we lived under constant threat of foreclosure. The farmhouse had few amenities. When I was very young, I slept on a bed that was made from an old door covered with a piece of foam. We never went hungry, but sometimes dinner was plain spaghetti sauce over noodles—for weeks on end.

My relationship to money remains infantile. Money scares the hell out of me. I don't know what to do with this stuff. Save it? How? Spend it? Sure! In Williston, I thought about money in the same stunted way as the dudes who got their first oil field paycheck and immediately bought new Ford F-150s. Money! It's great!

For Champ, Frank, Skip, Tammy, and me, our desires are so modest they could fit in a tin can. We aren't reaching for the brass ring; we're hustling just to get in the game. The only way to justify our experience is by saying we want money. Because worship of money is at the root of three magic words: The American Dream. I will hear those words spoken time after time in Williston. They are used to justify endurance of the harshest conditions. Climbing an oil rig in subzero temperatures without a properly fitted safety harness, I'll hear an oil field hand shout it: "This is the American Dream, motherfucker!" Falling down drunk in a whiskey-stained hotel bar, a nineteen-year-old on his way to bang it out with a bored hooker will slur it at me, "It's the American Dream, man." And in a parking lot, sleeping in a broken-down car, with gray at his temples and lines crowing from his eyes, an out-of-work construction worker will say it softly, a longing incantation: "This is *the American Dream.*"

Those three words stun me, confound me, anger me, make me laugh in irony and in joy, and, I admit it, they give me hope, too. Like all the other suckers. At Champ's, sleeping on the flophouse floor, the words resonate with me more than anywhere I have ever heard them, before or since. But what they really mean is: Money. And I don't know a damn thing about money. I just know I haven't had enough of it.

SEEKING

"Hello, ma'am. My name's Michael Patrick. I've got a résumé here. Was hoping I could fill out an application for you."

"Howdy there, sir. I saw that y'all are looking for more hands now that the rain's clearing up? Still sure is muddy out there, but the sun feels good. My name is Michael . . ."

"Yes, ma'am. I do have a place to live, and my driving record is clean as a whistle."

"No, sir. I've never been arrested."

"Well, that's all right, ma'am. If you don't mind, I'll just go ahead and fill it out so that you have it on file. And what's your name again?"

"No problem, sir. Here's my résumé here and here's the application itself. If you have any problem reading my chicken scratch, just gimme a call. Now what did you say your name was?"

"And, if I were to follow up with someone about the résumé, who would I ask to speak with?"

"Is it Ed who does the hiring? I think when I called, I may have talked to Ed."

"You have a great day now, Mary. Make sure to get out and enjoy a little of the sunshine!"

"All right, take care, Bob. Don't get swallowed by that stack of papers now. When you see Ed, let him know I'll be following up with y'all in a couple days!"

And out of the offices and into the sun and into the Chevy and down the road in circles around the hot, flat, tangled streets.

JESUS

Jesus is getting ready to leave for work. He is a big guy, well over six feet tall, with thick limbs and broad shoulders. He has a round, moon-shaped face and unsettled eyes. Frank sits on the couch looking at his phone.

I have just gotten home from my job hunt, another day of putting in applications and kissing ass. I'm taking a moment on the couch to wind down before I make myself dinner when Jesus tells Frank to stop taking his picture. He speaks in soft, warm tones, and I have to concentrate to understand what he is saying. Frank's bald dome takes the shape of surprise. "I'm not taking any pictures," he says.

"I don't want my picture on the internet," Jesus says.

"I'm not taking any pictures," says Frank, now concerned. He holds his phone out to Jesus. "Here," he says, "look."

Jesus refuses to look at the phone. "You're taking my photo so you can post it on the internet," he says.

I'm seated on the couch next to Frank. Frank looks at me for reassurance. I shrug, and he grows testy and stands up. "I'm not taking your picture," Frank snaps.

Jesus moves close to him. He is a head taller than Frank and twice as muscular. He lays a massive hand on Frank's arm. "Take your hands off me," Frank says, yanking his arm back. In deliberate silence, Jesus rests a hand again on Frank's shoulder. He leans forward until their heads are almost touching.

"I want you to stop taking my picture," Jesus tells him.

"Take your hands off me," says Frank, his voice rising into a high-pitched whine. "I don't want to get in any trouble." He pulls away from Jesus and moves quickly into the kitchen. Jesus moves after him. I stand up from the couch. I'm not sure what to do. If this escalates, do I call the police? I don't know where anyone else is.

"I don't want my photo on the internet." Jesus's voice stays low as he catches up with Frank. Frank holds his hands close to his body in a defensive posture, and Jesus again puts a meaty hand on Frank's arm and then another on his shoulder. They are silhouetted in the dark room by light through the screen door that leads to the backyard. My pulse quickens and my body tenses.

Suddenly the screen door swings open and slaps shut and Champ is in the kitchen standing squarely between the two men. "Do you wanna fucking move out?" he barks with rabid intensity. "Do you two wanna find another fucking place to live, because I'll throw you both out on your asses."

Jesus and Frank step back from each other, hobbled by Champ's vehemence. "No, no, no," they say.

"I don't want any of this shit between you two," Champ continues. "We gotta get along here." Frank tries to explain, and Champ cuts him off. "I don't care," he tells him flatly. "Now you two fucking shake hands, or I'm tossing you both the fuck out."

Frank and Jesus exchange a feeble handshake. "All right," says Champ. "Now I'm going to finish making some fucking burgers."

He heads back outside, and a strained peace follows. Frank leaves on a walk, and Jesus soon heads to work to stock shelves at Walmart.

A few days later, Champ learns that Jesus was fired from his previous job for harassing people for the same beef he has with Frank: pictures on the internet. "I don't need no fucking crazies around here," Champ says. He tells Jesus he has to move out, kindly giving him a refund for a full month of rent.

That weekend, Jesus leaves the flophouse in a broken-down car that he has no idea how to drive. I'm the only one home. He stands in the living room staring out the window, his face taut with worry.

"Do you know anything about cars?" he asks me.

"Not really," I tell him. "What's wrong?"

"Smoking, hissing, leaking," he says.

"I can't help you fix it, but I can show you how to drive it," I say.

He shakes his head. "I don't have time," he tells me.

He remains glued to the window, his neck thick with tension, his brain twisted over photos and the power of the internet. All his belongings are piled into two heavy-duty trash bags. He carries the two bags out onto the street and tosses them into his brand-new broken-down car. I watch through the window as he drives unsteadily into the bright North Dakota day.

I will see him one more time, almost a year later, walking down the street—stalking it, really—with a determined terror stretched across his round moon face.

CURSE OF THE
STARVING CLASS

The Williston Job Services Office doesn't allow guns, as the sign says, but another sign requests that entrants wipe their feet, and red mud is smeared all over the carpet. I've heard that this organization can help me find work, so I take a seat at a composite board desk and fill out forms on a desktop computer, carefully working my way through a long, extensive skills list. I click a box next to each skill I possess, which forces me to reckon with myself. I am handy, yes, but can I frame out a door? No, not really. I have painted the interiors of a few houses, but am I truly experienced at cut painting? Well, I guess not. I continue to scroll. The list goes on and on, page after page. When it is done, I blink at the computer screen, and wait for the results. How can it be, I wonder, after working literally dozens of jobs in my life, that I don't know how to do anything at all?

Of course, I do know how to do things. More than anything, I know how to make theater.

I went to school for acting. I was lucky to have found it as a kid.

My first day of high school I wore a black Jack Daniel's–branded cap I'd bought at the local carnival. My older brothers had both dropped out (well, Ryan dropped out, Matthew was expelled for bad behavior) and I entered high school determined to leave it as soon as I turned sixteen. I figured I'd work at a gas station until I was old enough to join the army. My father was a combat veteran, and I wanted, more than anything in the world, to be a US Army Airborne Ranger.

The man who walked out in front of me on the first day of drama class was thirty-three years old, with dark hair and bright eyes. It was his first day teaching, and he stood before the class with a slight stoop, wearing slacks and suspenders over a white shirt and tie. He clasped his hands together with a smack, then he began to speak.

"Do you feel nervous on your first day?" he asked. "It's my first day, and I feel pretty nervous. Is anybody a little scared? Raise your hands if you are. It's normal if you're scared. I'm kind of scared myself."

I glanced around the room uneasily, but I did not raise my hand. The man held his hand up and looked at each one of us in turn. My eyes were stuck on him. I had never heard a man talk directly and honestly, unashamedly about his feelings.

He must be gay, I thought. *Drama teacher, suspenders, feelings.* I'd be lying if I said that, at the time, this wasn't important to me. I was raised in the church and taught to despise gay people. This, despite the fact that my brother Ryan was clearly interested in boys. If my drama teacher had been gay, it would have been easy for me to view his sensitivity as a symptom of his gayness, and it would have been easy for me to dismiss him. As it so happened, he was straight. He also coached the baseball team. For me, a fourteen-year-old farm boy, to lay eyes on an athletic, unabashedly sensitive, emotionally astute, straight man? I can not express how grateful I am that he came into my life when he did. His name was Carl.

A few weeks into the semester, Carl told me I should audition for the fall play. On a whim, I agreed.

When my name was called to read a scene, my hands were trembling. I felt queasy, light-headed, and wobbly on my legs. I stepped onto the stage, I looked at the words on the page, and I said the lines.

I'd always felt apart from the other kids in school, as if I were marked, as if the thing that made me different—my broken family—was visible, a deformity I wore, and everyone watched me for it. But as I stood in front of my teacher and classmates in the auditorium, I felt, for the first time, free of their eyes, free of their judgment, free from my own constant worrying. It was easy. I suddenly belonged. And as I started to speak, to say the lines on the page in front of me, as I began to find wit in their clauses and music in their syntax, to imbue them with the roiling feelings that were bottled up inside me and begging for release, I felt the walls of the building disappear, the room itself disappeared, and, with the exception of my scene partners, all the people disappeared, too. I was blessed. My body had been stuck in a broke-down farmhouse with a shattered family and a rageaholic father. I'd been living inside my imagination for years. Without knowing it, I'd been practicing to do this my whole short life. Acting, for me, was the most natural thing in the world.

That night, I got down on my knees and prayed. I prayed harder than I'd ever prayed in my life, maybe harder than I ever would. The next day, posted on the door of the auditorium, was the cast list. My name was on it. This event changed the trajectory of my life.

Carl taught me the principles of acting and introduced me to great writing—to reading it, analyzing it, taking it apart like an engine and putting it back together. We studied Tennessee Williams, Oscar Wilde, Shakespeare, Eugene O'Neill, and Bertolt Brecht. I was especially enamored of O'Neill's early expressionist work like *The Hairy Ape*, as well as Brecht's sprawling episodic epic plays, broken up by songs and projections, riven with tension and dripping with empathy and condemnation for the morally and spiritually compromised. I internalized his idea of theater as a tool to make the strange familiar and the familiar strange, to force people to question their assumptions. This altered my perception of the world.

Carl gave me copies of plays like *American Buffalo* by David Mamet, where Don threatens Teach with a tire iron over a nickel that might be worth something. He gave me *Curse of the Starving Class* by Sam Shepard. It begins with Wesley, a teenager, cleaning up the screen door his father tore down the night before. The characters talk to an empty refrigerator. When Wesley's father comes home, he goes on a rant about castrating sheep and falls asleep on the kitchen table. I'd never read anything that so directly mirrored my own life. Shepard became one of my guiding lights.

Instead of dropping out of high school to join the military, I went to college on a scholarship. After college, I formed my own theater company, writing, directing, and acting in dozens of stage plays.

It is a weird résumé for a man applying to work in the oil field, I know, but it was in the theater that I found my first tribe. Athletes with bad knees, women who probably played with dolls all the way through high school, musical theater nerds, and ragged, damaged seekers like myself.

Through acting I found an outlet for my confusion at my upbringing, my shame at the abuse I'd witnessed, the fear I felt for my father, and for the deep, unrelenting, inarticulate pain that defined me. I was good at it, too. I worked in the theater throughout my twenties.

In my thirties, I could barely pay my bills. I needed dental work. I couldn't afford asthma medicine. I needed a doctor's care. The jobs I'd picked up for money over the years were labor jobs—a little carpentry, a little construction, a lot of stagehand work—but the pay sucked, and finally, just as I left my twenties behind, I injured my knee. I didn't have health insurance, and now I couldn't do physical labor.

I became a receptionist at a high-powered corporate law firm in the Financial District. For a year and a half, I spent many lunch breaks crying in a bathroom stall. But I was able to leverage that experience. With the help of a friend, I landed an administrative job at a creative agency in Midtown. I began in 2008, a week after the economic crash, in an office that shared a building with Lehman Brothers. As I rode the elevator up to the best-paying job of my life, dressed uncomfortably in an ill-fitting dress shirt and slacks, I shared it with workers who rode it down to the unemployment agencies. I'd be making $55,000 a year, which for me was a fantastic sum.

Once I became an office worker, theater moved from the center of my life to the outskirts. I found myself wondering what my life would have become had I never found acting. What if I'd dropped out of school like I had originally planned and become an Airborne Ranger?

DAY LABOR

At the bottom of town near the train station, in a low-lying building that looks like a general store in a Hollywood Western, sits Bakken Staffing, a day labor agency. Inside, men sit in folding chairs at mismatched tables piled with out-of-date *National Geographic* magazines. Big men, small men, fat men, and thin men, from their early twenties to, I'd guess, their late sixties, they wear unshaven faces, camo ball caps, and bored expressions. Two guys snore under a NO SLEEPING sign until a staff member, a woman named Katie with long, graying hair, wakes them up.

I begin filling out a stack of paperwork, and we watch a video on worker safety. One of the guys who had been sleeping nods off again. He has a big bald head and wears a dirty sweatshirt. Katie kicks his foot to wake him up.

"Is this your first time here?" she asks.

"Yes, ma'am," he says.

"Well, we're not impressed."

In a loud, firm voice, she gives us a rundown on how the company operates. "Once you finish your paperwork, you'll be given an

ID card. Every morning you want to work, you swipe your card. That way we know you're here, okay. The jobs pay between twelve and eighteen dollars an hour, and most of them are general labor. At the end of the day, Bakken will pay you out immediately, okay. We offer debit cards to all our workers, and we can put your money on the debit card within sixty seconds of you leaving."

She pauses to make eye contact with a few of the guys before she continues. "Now, I do the payouts at six p.m. If you're ever stuck late on a job, you can call me. If I can, I'll stay late for you, okay? Just let me know. If I can, I will."

I haven't worked an honest day in almost two months, and job hunting is starting to take a psychological toll. Somehow, the only thing I can compare it to is that feeling of breaking up with a girlfriend. I'm on a roller coaster of emotions, worn down and heavy-hearted but trying to stay positive. A few days of work could give me a boost of confidence. At the very least, it'll ease my bank account's hemorrhaging. I had $3,000 in the bank on June first. Two weeks later, I have just over half that.

I talk to the guy who keeps falling asleep. He'd hired on at Nabors as a water hauler, and he was doing well for a time, but they laid him off and kicked him out of his man camp a month ago. He moved into a hotel room and started looking for a new job, but his money ran out quicker than he thought it would—the cheapest motel in Williston still costs over $100 a night. And now here he is, in a dirty sweatshirt, falling asleep in day labor dispatch. He shakes his bald head in apology to Katie, rubbing his bewildered, unshaven face with a big, hammy hand. "I'm sorry," he says. "I'm homeless, and I'm just real tired."

"It's okay, honey," Katie tells him.

. . .

A FEW DAYS LATER, in the dim light of dawn, I return to Bakken Staffing and pour coffee into a Styrofoam cup, dump powdered creamer in, and take a seat among the guys. The men start chatting as the sun rises—a low, steady murmur building into a sound like bees buzzing. Somebody bitches about taxes. One guy says, "Try putting two pairs of socks under these boots." Another guy describes a car accident.

"Anybody got carpentry experience and a car?" one of the staff members at Bakken hollers, and I step up to the desk. They pair me with a carpenter and give us directions to a cabinet mill in Sidney, Montana. Sidney is an hour southwest of Williston, just over the Montana border. The town doesn't sit directly over the Bakken Formation like Williston. It isn't a boom town, exactly, but it suffers the residual effects of the boom: life adjacent to an earthquake.

The carpenter speaks with a slow, syrupy southern accent. He has the porous face of a drinker, long brown hair under a well-worn baseball cap, and an unruly beard. He's been in Williston a week, and this is the second job he's picked up through Bakken Staffing. As we ride to the mill, he tells me he went out the day before with a guy who had been up for three days, smoking meth. The guy was driving and kept falling asleep at the wheel. "I'm not supposed to drive," the carpenter tells me. "I drink too much. But I said, 'Pull the fuck over, buddy.' And drove us back to Bakken."

AT THE MILLWORKS, I get paired with a heavy-set, bearded Montana native named Jeff. We are building cabinets on an assembly line. "We once had over fifty workers," Jeff says, "but since the boom drove wages up, we haven't been able to keep anybody. Right now there's thirteen employees. Rent's gone up, too!"

Jeff shows me how he cut off the tip of his finger, and he tells me about a guy who got his hand stuck in the machinery. "They saved his fingers, but he can't feel them anymore," he says. He talks about a dude who got a board embedded in his stomach after it kicked back out of a saw. "He pulled the board out himself and somehow survived long enough for the EMTs to show up!" Jeff builds to the story of the time a tractor trailer rolled on top of one of the company's drivers. "They had to leave the guy under the wheel until the ambulance arrived. Somehow he lived, too, but his hip and leg and everything else is all screwed up. He still gets around, though," Jeff assures me.

"Did you hear about the local math teacher?" the carpenter asks me in a hushed voice on our lunch break. I shake my head. It would turn out to be a story I'd hear repeated many times. It seemed to haunt the oil boom.

"It is our worst local tragedy," Jeff tells me back on the line. Sherry Arnold—a forty-three-year-old algebra teacher and mother of three—disappeared on January 7, 2012, after leaving her home for a predawn morning run. A single shoe of hers was found along the running route that very day, but it wasn't until late March that her body was discovered in a shallow grave on a farmstead near Williston. Lester Vann Waters and Michael Keith Spell had been smoking crack ever since they'd left Parachute, Colorado. They had come to the area, they said, looking for oil field work. When they saw Sherry Arnold running along the road, the two men pulled her into their Ford Explorer and killed her. They drove to Williston, stopping at Walmart to buy a shovel and a package of bologna for sandwiches. Sherry Arnold was buried in a shelter belt, a line of trees used by farmers to shield crops from the wind, off the 1804 Highway. Three days later the men returned the shovel to Walmart for a refund.

On the drive back to Williston, the carpenter tells me he had come to the boom from Alabama with a buddy. They were sleeping in a car at Love's Travel Stops off Route 85.

I give him a ride to the truck stop, about ten miles north of downtown Williston. The parking lot is full of groups of guys standing by the vehicles they live out of, grilling burgers on the tailgates of their trucks, sipping on sodas and cans of beer wrapped in brown paper bags. A mixture of rock and roll and country music pours out of truck windows, the smell of hamburgers and cigarettes in the air. It does feel like a tailgate party, I think, watching the young men goof off in the never-ending North Dakota evening. I drop the carpenter there and drive back to the flop.

I will see the carpenter once more, a few weeks later while shooting pool with the welder's assistant I met my first night in town. The carpenter meets us after his shift, his clothes covered in a thick black tar. "Another Bakken Staffing job," he says, shaking his head in dismay. He'd spent the day applying some sort of sealant to concrete walls in a basement. "My arms probably smell like gasoline because I had to use that to get the tar off my arms," he complains.

His friend with the car had "started freaking out"; he was skipping town that night. So the carpenter planned to set up a tent at an RV park a mile north of the truck stop. "I'm just worried about my tools," he tells us. He buys a beer and drains half of it in a single swallow.

"Everybody has to get through this part," the welder's assistant tells the carpenter.

"I know," the carpenter says. "I ain't leaving. I can't do any more through Bakken Staffing, though; the jobs are just too bad. Maybe I'll pick up some construction work after all."

"That would be stupid," the welder tells him bluntly. "You'll be working just as hard as anybody else but you'll be making less money. And you'll be stuck in Williston."

The carpenter shakes his head. "I wanna get an oil field job," he says, "but I'm worried about my license." He'd gotten four DWIs before moving to Williston.

"Every place I've applied has asked about my driving record," I tell him. "One spot ran a report as soon as I put in my application."

He grows even more downtrodden at the news, takes another gulp of beer. "It seems like all anybody talks about up here is work," he grumbles, wiping his mouth.

"That is all anybody talks about," I tell him.

When he heads out the door, we watch him go. "I wish him the best," the welder's assistant says, "but he don't look much like the type to make it."

"No," I say, "he don't." I'll never see him again.

BROOKLYNITES

Chicken and onions sizzle in a pan. The smell of garlic and cinnamon drifts out from the kitchen. Daniel and Gabriel—two of the Jamaicans—are cooking. Daniel, stirring the vegetables with a wooden spoon, has the lean, bandy build of a professional soccer player. He is very dark-skinned, almost purple, and gorgeous, and he knows it. Gabriel is taller and thicker with a serious face. He watches over Daniel's shoulder with a circumspect expression. Both men are polite and guarded. They live apart from the others in the flop. They have their own room, they work third shift, and they are black.

"So, you guys are exchange students?"

Gabriel's face remains placid, but I see a subtle expression of surprise in his eyes. "No, no, no," he says, "we've both finished school."

"Champ told me you were students," I tell him. "I thought that was weird."

"No, no, no," Gabriel tells me. "We are here for a few months, then we head back home. We work at the gas station. Sometimes we work at a bodega in Bed-Stuy," he says.

"Bed-Stuy? In Brooklyn?" I say. "I'm from Brooklyn."

Gabriel's eyes light up, and his features soften. He breaks into a smile that dazzles. "Mikey! My man!" he says. We laugh and clasp hands. It is as if a clean cool breeze blows through the house. Our words come out in a tumble as we talk about the city. Gabriel teaches me to say "What's up?" in Jamaican. It sounds like "Wag wan?" I repeat after him a few times before I get it. I'm completely marble mouthed.

"How do you say, 'I'm good'?" I ask him. "Like I'm doing good?"

Gabriel pauses, looks at Daniel, then back at me and shrugs. "Me good," he says. The three of us share a robust laugh. "Jamaica doesn't really have its own language. It's pidgin," he explains. "It's a, uh, broken English."

We share stories of New York, New York, spending some time simply naming streets and subway stops we both know. It is incredibly comforting, and I realize that I like the Jamaicans in the house, especially Gabriel, immensely.

IN RURAL MARYLAND, I grew up almost exclusively around white people. In school, I occasionally heard racist jokes, and some of the kids and parents I knew were outright white supremacists. There was a certain amount of Klan activity where I grew up, too. I have one memory, when I was very young, of the Ku Klux Klan demonstrating in the streets of Damascus. It was summer, my mom was driving, and we were stuck at a traffic light, when a tall man in a white hooded robe eerily, silently, reached into the car and handed my mother a pamphlet. She accepted it, but quickly rolled up the window. When I was a teenager, the Grand Dragon of the KKK was a contestant in the Great Frederick Fair's demolition derby. He'd painted his car with Klan signs and wore his stupid little hood and robes as he crashed around into other cars. This was around the same time, at the same fair, that the Pagans and the Hells Angels got in a turf war and a couple guys got cut up pretty bad, but nobody seemed to think much of it. The stories were considered oddities more than anything else. Like most Americans my age, I'd been taught, insanely, that the civil rights movement had ended in the sixties. It wasn't until much later in life that I began to contemplate how ignorant I was.

Not long after leaving the farm, I remember walking down Howard Street in Baltimore through throngs of shoppers, working folks and hustlers. I tried not to gawk, but I'd never been surrounded by black people. I'd never experienced black culture. I'd never, in any setting in my life, been a minority. It was frankly scary to be so suddenly apart, but it was also wonderful to feel my small world expanding in front of me with each step.

I lived in Baltimore for a decade before moving briefly to Los Angeles and then to Brooklyn. My first two years in New York, I lived in Sunset Park—one of the most diverse neighborhoods in the world. I was surrounded by colors and languages. I shared an apartment in a Polish-owned building with a Korean American, a German, and a Colombian. I bought tacos from El Salvadorians, worked with Dominicans, African Americans, and white folks from the Bronx. A Middle Eastern family at the neighborhood dollar store one time gifted me a plate overflowing with food, jasmine rice and spiced chicken. I spent seven of the best years of my life in Brooklyn.

Williston on the other hand was incredibly homogenous. It had been more so before the boom. But even with a tripled population (some estimated it quintupled), whites made up the vast majority of migrant workers descending on the town. These were the type of hardworking, blue-collar, truck-driving, jeans-wearing dudes I'd grown up with in central Maryland. It was a group of people that, even though in many ways I was one of them, I didn't feel completely comfortable around. I'd left it behind.

I never spoke directly with Gabriel or Daniel about the racism they encountered in town. In some ways, Gabriel's guarded expression told its own story. I can only imagine what it was like for them.

That evening, they introduce me to Cholly, another tall, thin Jamaican with a big, bashful smile. With Jesus gone, there is a mattress up for rent. Cholly talks to Champ, pays him in cash, and moves in that night, joining Frank, Jesse, and me in the living room.

SEARCHING

The walk from my SUV to the front door of each potential job becomes a ritual in shedding my fear and insecurity. With each step, I bring breath into my lungs and straighten my back. I swing my arms, steady my eyes, and pull my focus toward a future with a start date.

When I leave, I deflate like a balloon, slump back behind the driver's wheel to scratch another name off a long list, and turn the key toward the next stop.

At the next place, with each crunch of gravel underfoot as I move toward the doors of another job prospect, I will strengthen again. So when I swing open that front door, a hard-fought trace of a smile will grace my lips, and my voice will ring clear, "How you doing, y'all? Another day in the dust bowl, huh? I hear you all may be looking for a few sturdy oil field hands . . ."

THE LAST SHOT

I am sitting at the kitchen table alone when the back door swings open and in one single uninterrupted movement Champ enters, opens a bottle of beer, drops a plastic bag on the table, takes a pull from his beer, and sits down across from me.

He opens up the bag, and it is full of cash. He just sold his Harley-Davidson. "Whaddaya say we get a big rock of cocaine, a motel room, and a couple hookers?" he says.

"Don't tease me like that," I tell him.

He snorts at my smart-ass response, surprised he didn't shock me. Then he tells me about the time he was paid for work with a huge crack rock. He smoked for sixteen hours straight and put himself in the hospital. "The doctors were cool but, man, the nurses busted my balls."

Champ sits the same way he stands, with his feet splayed and his shoulders squared. He's in a perpetual wrestling stance, always at the ready, his broad, sunburned face turned toward the world. He talks about his job, hauling oil for a company called Diamondback Trucking. "It takes some real skill to navigate the fucked-up, cobbled-together infrastructure out here in a tanker filled with ten thousand gallons of sweet crude," Champ says, pulling on his beer and running a hand through his uncombed hair. "It's not like haul-ing pipe or water or sand. Pays a lot better than that. I'm doing five shipments a day right now and money is rolling in." His eyes move as he talks, counting the loads, counting the gallons, counting the money. "You oughta apply at Diamondback," he tells me. "Swampers start off making eighteen to twenty dollars an hour."

"Swampers?" I ask. "What's the job?"

"Being a nigger," he tells me. The word drops with a thud. He finishes his beer and goes upstairs to bed.

ON ANOTHER NIGHT, we stand in the garage behind the house, Champ holding an M14 assault rifle in his hands, displaying it for a potential buyer, a trucker he works with at Diamondback. The driver inspects the barrel of the gun; he likes that it is attachment ready. A cool breeze passes through the garage.

"This gun is powerful enough with regular ammo. It'll go through a human being and through anybody standing behind them." Champ gestures to a table full of ammunition, including boxes of "zombie stoppers," soft-tipped bullets that can blow huge holes through a human body. "We're ready for the apocalypse around here," he cracks.

He hands the rifle to the potential buyer, saying, "Totally un-traceable. I bought this with cash at a Montana gun show. If Obama ever tries to round these babies up, they'll never find this one." The trucker handles the gun with practiced ease, cradling it in the crook of his elbow. He turns it over and admires it.

Champ had bought the rifle following the massacre at Sandy Hook. "All the guns around here went quick," he tells me. "Weapons brokers were cleaned out. That sent Washington a fucking message."

He shows me a handgun he wants me to buy. It is a .45 caliber automatic. He pushes the clip into the handle of the gun and passes it to me. I sight it and look it over. It feels heavy in my hand.

"And look at the barrel," Champ points out. "You can attach a silencer here."

There is an odd thrill in holding it. I haven't handled a pistol since I was a teenager. I imagine having this .45 in my luggage in the living room of the flophouse. "Too much gun for me," I tell Champ. "I don't know what I'd do with this thing."

The trucker says he likes what he sees. Champ makes an offer, and they haggle for a minute, quickly settling on $1,600 for the rifle and the ammunition. Then they shake hands. It's a deal.

SUMMER EVENINGS in North Dakota are beautiful. The sun hangs on the horizon seemingly for hours, projecting a hot yellow light across the plains that creates an ever-changing skyscape. Even surrounded by houses, the colors of the evening twist and turn slowly before sliding into the purple night. The dry heat cools gradually. The clean air feels good in the lungs.

I sit on the back patio and work through songs on my guitar. I've played guitar and performed in bands since I was a teenager. My high school drama teacher, Carl, played guitar, and he turned me on to it. Since then I've met most of my friends through music. Music sits at the center of my life. I'm a folk singer, but I've shared bills with some prominent indie rock bands and I've opened up for Ramblin' Jack Elliott, America's last great true folk singer.

The guitar is the most useful tool I have ever owned. It accompanies the voice, tells stories, and transports souls. It is a home you can carry with you in a case. Open it up, strum a chord, and you are exactly where you are supposed to be. There's a reason so many travelers are guitar players, and there's a reason bands hit the road. It is a refugee's instrument because it is a time machine, a storage house for memories that also makes people feel completely present, bracingly in the moment. Anywhere on earth, the sound of a strummed six string provides shelter for our souls.

Champ joins me with his guitar. He owns an expensive Martin. It has a deep rut across the body from a moment when, in a pique of anger, he threw it across his bedroom. Despite the angry tattoo, the instrument plays well. Usually, Champ will attempt a solo and then say, "Well, I fucking suck," and go back inside. But sometimes we just sit around holding our instruments and talking.

On this night, Champ wants to sell his guitar. I'm not interested in it, though. I can't afford another guitar. He presses me on it. "I wanna sell all the extra stuff I have," he says. "Outta all my friends, I'm the fuckup, the loser. This is my last shot."

It is date night, he tells me. He should be spending it with Rickie. "She's unhappy with the situation in the house," he says sarcastically. "She doesn't like having so many people around, but she doesn't pay any fucking rent, either. And man, she was so drunk yesterday. I asked her to do just a couple things, you know. Could you either clean the bedroom or clean out the fridge? She didn't do a fucking thing. I get home, and she's fucking annihilated. She gets all clingy, too. 'I love you, I love you, I love you.' I hate that shit."

He thinks about it a moment, drawing his bottom lip up and moving his head from side to side. "But she did stick it out with me for a year in that trailer," he says.

"That's worth something," I say.

"Yeah. I guess it is."

The trailer froze up in the winter, and they didn't have hot water. Champ would get home from work, turn on the space heaters, and then drive to the rec center for a shower and return to the trailer once things warmed up. After one winter, he decided he couldn't take it anymore. So he sold the trailer and bought the house. He wants to get it paid off as quickly as possible. That is why I am here. That is why Skip and his mom, Tammy, bald Frank and Jesse, Gabriel, Daniel, and Cholly are here. That is why the living room is full of mattresses, why the fridge is clogged with food, why the bathroom floor is always wet and the washer always full.

In a normal town, people follow the infrastructure. If you move to Chicago, for instance, there are apartments waiting to be rented; in a boom town, the infrastructure follows the people.

Champ had bought his house right before prices became impossible. Now, he's racing to pay it off. "It ain't going to last forever," he tells me. "I'm gonna get mine, pay off this place, and unload it before the whole fucking thing goes belly-up."

For the time being, though, the house is full of people, and Rickie and Champ are having problems.

"She's got a head shrinker, Rickie does," he says, "and she tells this broad how she's not happy with the living arrangement here, having all these people around. And her fucking head shrinker tells her that she should give me an ultimatum. She tells me this, Rickie does."

Champ looks at me with his eyebrows raised, his eyes wide, and his brow creased. His spiky red hair frames his round red face. "I said, 'You do not want to fucking do that. You will not come out on the right side of that equation.'" We sit for a few moments in silence, sipping our beers. "I got to make good," says Champ. "This is my last shot."

I feel a similar desperate hunger gnawing at my bones. It isn't that this boom is the last opportunity I'll ever have, exactly. But I am taking a big swing, the biggest I've ever taken, and, if I miss, well . . . I can't miss. I just can't.

CHAMP HEADS INSIDE and I sit alone with my thoughts. The motion sensor in the backyard keeps turning the lights off, and I have to stand up periodically and wave my arms to get the lights back on. Frank walks out of the house in his Pepsi uniform. He takes the cap off his bald head and sits down with me.

Much of what Frank tells me would turn out to be a lie. The day I met him, for instance, he said that he drove a truck for Pepsi. He said it with a certain amount of pride—he was a *truck driver*. It wasn't true. He worked in the warehouse loading boxes onto trucks; he didn't know how to drive one.

On this evening, however, I think what Frank says is true: He tells me he is twenty-six years old. With his shiny bald head and creased brow, he could pass for forty. He grew up outside of Minneapolis, he says, and he asks me about New York. Was I there when the towers fell?

"No," I tell him, "I was there the week before and then a couple weeks after."

"My brother signed up after 9/11," he tells me. "He was killed in Iraq."

"Oh."

"My mother had a nervous breakdown and died not long after that," he says. With a small inheritance, he moved from Minnesota to the Philippines, where he met his wife. They lived there for seven years until his inheritance ran out. Then he returned to the States.

He tells me that his brother's widow still lives outside Minneapolis with their kids, his nieces and nephews. After leaving his own wife and child in the Philippines, he traveled back to Minnesota to visit his brother's family before coming to Williston for work.

"I came out here on a bus in January with sixty dollars in my pocket. It was cold," he says.

"Do you work in the morning?" I ask him.

"Yeah." His voice is high-pitched, nasal, and whiny. He sighs, leans his bald head in, his full, thick lips and wide, pale face. "I gotta go lay down," he says.

SAKAKAWEA

One bright North Dakota day, on a break from my job search, I take a lonesome walk along Lake Sakakawea, a man-made lake in the central part of the state. In a cruel irony that epitomizes the federal government's centuries-long mistreatment of Native Americans, Lake Sakakawea was built on top of the ancestral home of the Mandan, Hidatsa, and Arikara nations as part of a network of dams created to prevent flooding primarily in white communities in Iowa and Nebraska. Creation of the lake in the 1950s, on some of the most fertile bottomland in North America, uprooted thousands of tribal members and led to the destruction of an incredibly vibrant Native community. As Paul VanDevelder details in the book *Coyote Warrior*, in some cases, houses were moved without any notice at all. He writes of a pair of newlyweds who feel their house tilt beneath them during dinner. The husband runs outside to find a construction crew in his yard. "We're moving your house, Mister. Where do you want us to put it?" they asked. In other instances, army veterans returned home to find the houses they'd grown up in had disappeared, been moved to a different town. When the flooding took place, a group of students had to pilot boats through the hallways of their high school in order to retrieve the school's athletic trophies. Many tribal members ended up in the town of Parshall where, cut off from the hunting and farming that had sustained them for generations, alcoholism and diabetes ran rampant and racial resentments with white locals sometimes boiled over into violence.

North Dakota is still home to myriad Native peoples as well as five different reservations. Fort Berthold, which encircles much of the western part of Lake Sakakawea, comprises what remains of the ancestral homeland of the Mandan, Hidatsa, and Arikara, now known as the Three Affiliated Tribes. Spirit Lake Nation is organized around the town of Fort Totten, on the shores of what is now called Devils Lake. A sliver of the Lake Traverse Reservation, home of the Sisseton Wahpeton Oyate, crosses from South Dakota into southeastern North Dakota. The Standing Rock Reservation, which would soon become ground zero for protests against the Keystone Pipeline, is home of the resettled survivors of Sitting Bull's Sioux warriors. The Turtle Mountain Indian Reservation, home to the Turtle Mountain Chippewa, sits in the extreme northern part of central North Dakota, bordering Canada. While the tribes all share certain things in common, their heritage and histories are as distinct and disparate as those of any group of European nations.

Williston itself is located smack dab in the middle of the Trenton Indian Service Area, a murky legal designation that ties the region to the Turtle Mountain band. Most of the Native guys I would work with were Chippewa, also called Ojibwe. The majority of them lived in Trenton, the heart of the Service Area.

Walking around Lake Sakakawea, I notice a group of vultures circling the sky above me. I expect to come upon a dead carcass of some kind—a deer maybe, a badger or raccoon. Behind each bend of the rocky shore, through the tall grass and tough shrubbery, I continue forward but I smell no rot and find no carcass. I walk some distance before realizing the vultures are, in fact, circling me. *What country*, I think. I pick up a rock, draw my arm back, and aim for the birds.

Lewis and Clark, on their journey across the continent, spent more time on North Dakota soil than anywhere else in the country. Their first day in what is now the state of North Dakota was spent administering seventy-five lashes to a recalcitrant Corps of Discovery member named John Newman for "uttered repeated expressions of a highly criminal and mutinous nature." This display of torture led to a cry of dismay from a Native guide, an Arikara leader called Arketarnarshar. Arketarnarshar, according to Clark's journals, explained that the Arikara "never whipped even their children." The leader hinted strongly, of Newman, that Clark should just kill him.

The Corps of Discovery did not plan on staying so long in modern-day North Dakota. After experiencing several "verry Cold" nights and weathering their first snowfall on the plains, Lewis and Clark wisely decided to hunker down. Fort Mandan, named for their tribal neighbors, was constructed at the confluence of the Missouri and Knife rivers in the central part of the state. The Corps did not move from there for five months. Were it not for the kindness of the Mandan, and their willingness to trade provisions they'd saved for winter, the Corps of Discovery most likely would have starved to death.

Standing today in the more than 70,000 square miles of prairie grass that makes up the state can feel more hole than doughnut, but North Dakota is the geographical center of the North American continent, roughly 1,500 miles from both the Atlantic and Pacific oceans, the Gulf of Mexico, and the Arctic Archipelago.

The Mandan Tribe called their home The Heart of the World. They were a primarily sedentary people, living in villages made of large earthen lodges just outside of the river's floodplain. They farmed the soil of the river valley and engaged in annual buffalo hunts on the prairie, but primarily the Mandan were traders. As Paul VanDevelder writes, "the Heart River Villages were the commercial hub of a trade and distribution network that linked the Aztec and Toltec cultures of Mesoamerica to the Cree of northern Quebec. Comanche of the Southwest brought Arab stallions and Spanish knives to trade with Hudson Bay Assiniboin, who bartered English flintlocks, gunpowder and textiles."

Mandan life was organized to a certain extent around a summer festival called the Okipa Ceremony. Over the course of four days and nights, the tribe participated in, among other things, an extremely strenuous ritual enactment of the creation myth: Lone Man's arrival at the Heart of the World, his outwitting of Speckled Eagle at Dog Den Butte, and the creation of the buffalo-hide-covered turtle drums that end famine and restore balance. During the ceremony, the Mandan sang, chanted, danced, and prayed, dressing as, and mimicking the movements and sounds of, the animals of the plains. They engaged in burlesque-like sex games with a figure called the Foolish One. Young men fasted and mutilated each other with knives and hooks, leather thongs, wooden splints, and hatchets. On the final evening of the festival, the young women had sex with the old men of the tribe, in order to transfer the *xo'pini*—sometimes translated as "medicine"—to their husbands. To the white men who witnessed and wrote about it, as well as to most modern people, the Okipa defy an easy interpretation. Even reading about it, the ceremony unfolds like a strange dream. To experience the Okipa could only have been terrifying, awe-inspiring, and joyous.

The ceremony was presented by a single man. As Elizabeth Fenn writes in *Encounters at the Heart of the World*, "In present-day parlance, the Okipa Maker was the man who 'sponsored' or 'gave' the event, which was truly a gift to his people." To become Okipa Maker, over the course of a year, this man had to accomplish many difficult feats. He had to "dream of buffaloes singing the Okipa songs." He had to gather hundreds of items of ceremonial clothing, and he had to collect enough food to feed the entire village for four days. He also had to supplicate himself, soliciting "help from his wives, his mothers and sisters, his clan, and even his father's clan."

Only a capable man with rich social connections could hope to become an Okipa Maker. By sponsoring the festival, the Okipa Maker would, very often, impoverish himself. From the most celebrated man in the tribe, he'd quickly become it's most needy. In this way, the ceremony was a physical manifestation of the social fabric of the tribe, reinforcing a complex web of domestic and religious connections. It was based on an idea of radical sharing so far from our bifurcated understanding of capitalism or communism—or even charity and greed—as to seem heretical.

It was at Fort Mandan that Lewis and Clark met the remarkable young Shoshone-born woman who would travel with them and act, along with her husband, as an interpreter for the Corps. Sakakawea, whose name means "Bird Woman," could speak several Native American languages. Her brief biography provides a capsule look at the life of the Plains Indians at that time. Born Shoshone, Sakakawea had been kidnapped by a Hidatsa raiding party that attacked her camp and killed several of her family members. Held in captivity by the Hidatsa, she was later sold, or possibly lost in a game of gambling, at around the age of thirteen, to a French-Canadian fur trader. When Lewis and Clark met her, the teenage girl was pregnant with his child. Not much else is truly known of Saka-kawea but that she acted bravely during a boating accident. In later years, William Clark would all but formally adopt her son, Jean-Baptiste, and pay for his education.

In one of the few written accounts of Sakakawea found in Lewis's journal, when the Corps of Discovery was wintering at Fort Clatsop in modern-day Oregon, within spitting distance of the Pacific Ocean, the young woman demanded she be allowed to see a whale stranded on its beaches and view the ocean. Lewis recorded, "she observed that she had traveled a long way with us to see the great waters, and that now that monstrous fish was also to be seen, she thought it very hard she could not be permitted to see either (she had never yet been to the Ocean)." Her request was granted.

I like to picture Sakakawea there, solitary in the company of soldiers, with them but apart from them, her face pressed west against the cold Pacific wind—colliding with it—as salty waves crash over the sand at her feet. In a short life filled with brutal violence, she had been stolen, bought and sold, turned into property. Yet there, having traveled over a thousand miles, at the edge of a continent, she demanded agency. In a life remarkable for its singularity, the Shoshone-born Bird Woman created, from the dark materials of cruel existence, a beautiful moment belonging to her and her alone.

Historian Clay Jenkins would write, "For Lewis and Clark, North Dakota was the transition zone between the familiar and the unfamiliar, between lands that had already been mapped, named, and described, and *terra incognita*." North Dakota felt like that to me, too. When I first arrived in the western part of the state, I camped for several days in Theodore Roosevelt National Park. I hiked through the badlands and watched a herd of elk race across the valley floor.

I stumbled upon several small groups of buffalo, warily steering clear of them and nearly getting lost in the maze of burned red crags. I wandered freely among a colony of chirping prairie dogs. One night coyotes woke me as they raced through my campsite barking and snapping. One morning I saw a beautiful yellow bird I could not identify. There was so much I couldn't identify. So much I did not know. The eternal height of the sky, the mighty sweep of the horizon, the very length of the day: there was a newness, a strangeness to this world I encountered. With the people, I felt as alienated as an expatriate. I was wary, watchful, quiet. The boom was its own nation, with its own laws, ceremonies, and customs. Surrounded by all that mystery, I could sense the unknown awakening inside me as well. Questions I thought I'd answered long ago began to blossom in my gut: Who am I? What am I doing here? Why does any of this matter?

HIRED

Y ou got a record?" he asks me.

"Uh, no, sir."

"Can you drive? You got a good driving record?"

"Yes, sir. I can drive. My record is clean."

"You on the run? Nobody's looking for you?"

"No, sir. Nobody's looking for me."

"Nobody's looking for you?"

"No, sir. I'm not in any trouble."

"You got a place to live?"

"Yes, sir. I do. I moved into a spot a couple weeks ago."

"You sure you don't have a record?"

"Yes, sir. I'm positive."

"I need a swamper. Come on back. We'll get you started on the paperwork." I follow him back into a small cluttered, wood-paneled office. "It's that simple, huh?"

"I guess so," I say.

"I guess all this traffic isn't anything to you? Why'd you move out of New York?"

"I wanted better work, more money. I didn't really fit in there."

"What kind of work have you done?"

"Stagehand work, setting up concerts and events. Some rough carpentry, construction stuff."

"Are you in any trouble?"

"No, sir. I stay out of trouble."

"Are you afraid of heights?"

"Well, I don't love them, but I don't hate them, either, I guess."

"Are you reliable?"

"Yes, sir. I think so."

"Do you like to work?"

"Me?" I smile. "I love to work."

I sit in a rickety desk chair. The job pays $21.25 an hour. I ask about overtime. "There's work to do," is all he says. He hands me a pen, and I make my mark. It is June 20, 2013. Less than three weeks after arriving in America's Boomtown, I sign on as an oil field hand at Diamondback Trucking.

I'd been to well over a dozen companies all over Williston. Several had proven to be dead ends, but I was on a first-name basis with three different receptionists. Earlier that very day, I'd talked on the phone with a tool pusher who was hiring roustabouts. Diamondback Trucking may have been the place where Champ worked, but I hadn't dropped his name. I'd later learn they had a list of names and references five pages long. I got the job because I walked in a day after somebody else quit.

THE FINALS

While I was hiring on, Jesse and Frank were tramping about town trying to find a second job, or some financial relief, for Jesse. Jesse had been working with Frank at Pepsi but he wouldn't receive a paycheck until after the first of the month, so he didn't have rent for Champ.

That morning, I had dropped the two of them off at the house of a guy who needed some drywall work done, but when they got there they learned the work was finished. They then headed to Bakken Staffing to sign up for day labor, but they missed the morning safety video, so Jesse couldn't get put on payroll. After that, they walked to the Salvation Army, which had been giving food and, in some cases, financial help to migrant workers. But the process took a couple weeks; it would be no help. From there Frank guided the two of them to a Catholic church to speak with the priest. Before leaving the church, Frank and Jesse paused in the sanctuary. They knelt and prayed.

That afternoon, I sit with them in the sun-dappled backyard of the flop and eat burgers wet with ketchup and mustard off paper plates. "It was a little much for me," says Jesse in his syrupy Georgian drawl. His lanky frame draped over a patio chair, he wipes a smudge of mustard off his oversized T-shirt. "I mean, I've seen Catholic churches on TV and stuff, but I always thought they were exaggerating. I guess they're not." He looks at me with a somewhat bemused expression. He hasn't shaved his head in a couple weeks, and his mouse-brown hair has grown in unevenly.

At the church, the priest did not give Jesse any money, but he gave Frank a rosary. Frank for once is not wearing a Pepsi uniform. He's dressed in a green Adidas track suit, well cared for, but slightly faded and frayed at the cuffs. He pulls his shirt up and shows me the rosary. It is wooden, ornate, and elegant. Frank looks at me unblinking. His blond eyebrows disappear on the canvas of his pale skin. The sun bounces off his bald dome. The solemnity of Frank's action and the beauty of the rosary stir something in me that I haven't felt in some time.

Jesse massages the dirty bandage on his busted hand. This is his second stint in Williston. He had moved to the Bakken over a year previous and landed a roustabout job at Wisco, working pumping units. It was a hammer-swinging job, true oil field work. He lived in a man camp and made good money. He worked long hours, but there was downtime, too. He tells the story of one day waiting five hours for some equipment to arrive.

"I took a five-hour nap," Jesse says. "On the clock. Made what? A hundred and twenty dollars for sleeping. It was a good nap."

But the company wasn't friendly. For months no one so much as grunted hello to him. He was lonely. He missed his daughter, only a few months old, and his nineteen-year-old girlfriend, both 2,000 miles away. Winter came, and he spent three days on Amtrak traveling across the country, seeing none of it, returning to western Georgia to spend Christmas with his family.

"Wisco fired me, and I looked for work in Georgia but didn't find nothing," he says. "My girlfriend and her mom work like two nights at the Best Western. I'm the bad guy for coming up here, but what am I supposed to do? Stay there and work two nights at the Best Western every week? Making seven dollars an hour?"

His savings ran out, and he was right back where he started. Then his father called him. "He was living up here. He said, 'Come back out, I got a trailer.' I spent five days on the Greyhound. I get to the trailer and all his stuff's outside on the ground. He says he's on a fishing trip. Turns out he moved back to Georgia." The kid pauses to spit dip into a bottle. He shakes his head, "I don't know why you'd do that to somebody. But that's just my dad."

I feel good having finally landed a job, but I keep my excitement muted. I don't want to gloat. We head inside to watch the deciding game of the NBA Finals. There's nothing else to do. We've been watching the games all week.

Frank and I sit on the couch, and Jesse stretches out on his air mattress on the floor. His feet hang off it. We're gathered around the TV like previous generations of men sat around fires. Frank puts on a pair of safety glasses to watch TV. They are his only pair of prescription eyewear, and he looks incredibly ridiculous wearing safety goggles in an Adidas track suit with his legs spread wide like a gangster. He pours from a bottle of spiced rum into a pair of plastic cups full of Pepsi. "No Coke in this house," Frank says, pointing to his work cap sitting on a chair. He and Jesse slug down their rum and Pepsi.

Frank rubs his bald head. He tells us that LeBron James tried hair implants, but they didn't work. He says Bruce Willis also tried hair implants, but they didn't work. Frank says that he wants to get hair implants, but he is worried they won't work. "In 2017," Frank tells us, "they'll be able to stop the gene that causes balding. I'm going to get that."

"Are you kidding?" asks Jesse. "You can't change your genes after you're born."

They debate this. Jesse asks Frank if he knew his dad.

"No, I never knew my dad," Frank says. "My mom never told us about him."

"Was he bald?" asks Jesse.

"I don't know," Frank says. "I never knew him."

"Oh," Jesse says, and pauses before continuing, "my brother gets out of jail in November. Been in for three years."

"Did he get hair implants?" I ask.

"No."

"What did he do?"

"He robbed a house. Robbed my ex-girlfriend's house, actually. And he was on probation. He got three years."

Jesse hasn't seen his brother since the day he entered prison, and he isn't looking forward to seeing him again. "I know he'll come asking for money," Jesse says. "And he thinks he's all gangsta and shit." Right before he was locked up, Jesse had tattooed the word BLOOD in giant capital letters across his brother's stomach.

"Jesus Christ," I say, keeping the rest of my thoughts to myself.

Again, we fall into silence. We stare at the screen as the athletes shoot hoops.

Skip walks through the room, his shoulder and neck twitching as he crosses in front of the TV. He loudly proclaims his dislike for basketball and walks up the stairs. A short while later, he returns to sit on the couch between me and Frank and talk over the game. A commercial comes on for a movie about the end of the world.

"All these doomsday movies," says Jesse, "they start to get on my nerves."

"Why?" asks Skip. "Are you worried it's gonna happen? It is gonna happen, but it doesn't matter if you're saved."

"You know what I don't get about the Bible?" Jesse says. "Dinosaurs. It says God created man and then the next day he created animals. But how'd he do that if the dinosaurs were around for millions of years before that?"

Skip tells Jesse that he needs to have faith. Jesse says he would, but the dinosaur thing hangs him up. "You gotta have faith," Skip tells him. "You just have got to have faith." Jesse isn't so sure, but Skip continues sagely, "The three most important things in life," he tells us, "are money, pussy, and praising the Lord!"

Frank takes his goggles off and looks at Skip. "Money, pussy, and praising the Lord!" Skip repeats, joyfully.

Frank puts his goggles back on and resumes watching TV. He begins growing more and more serious. He bet $20 on the Miami Heat, and the game is on the line. In the final minute, the Heat pull away from the Spurs and win to become NBA champions for the second year in a row. As soon as the game ends, Frank gets on the phone.

"I want my money, bitch," Frank screams. "You better get me my twenty fucking dollars, bitch!" There is a pause as his coworker tells Frank he'll settle up with him on payday. "No fucking way, bitch!" Frank screams. "I want my fucking money tomorrow fucking morning! I want my fucking twenty dollars, and you better get it to me, or I'll kick your fucking ass!"

The coworker hangs up. Frank can't believe his gall. Outraged, he calls him back. The guy hangs up again. Frank calls again, and when his coworker answers, Frank hangs up on him. "Bitch," Frank says. Skip gets up from the couch, says "good night," and disappears upstairs.

Frank and Jesse finish their bottle of rum and start in on a bottle of whiskey. I tell Jesse he should lay off it, but he ignores me. He is so drunk he looks like he could fall down while sitting. Frank grows suddenly maudlin and starts talking about his brother who was killed in Iraq. We ignore him. But then Frank's coworker sends him a text calling Frank's wife a whore, and Frank decides he is going to walk across town to beat him up. I think that is a great idea. I'm sick of Frank. If things really work out, I figure, he might get arrested. I encourage him to find the guy and whip his ass. "Get that twenty bucks, bro," I say.

Of course, he doesn't walk across town. A different coworker shows up at the house. He has a look on his face that I associate with teenagers who don't know if they actually like their friends, but hang out because no one else is around and they want some trouble.

Frank heads outside with this guy and Jesse stumbles out the door behind them. They all pull their shirts off in the front yard and start tackling each other, forcing each other to the ground and choking each other out.

I don't want anything to do with it, but morbid curiosity gets the best of me. I look out the front door to see what is going on. Frank is facedown in the mud getting his ass kicked by Jesse and then by the other guy. I am surprised the neighbors aren't calling the cops. The yelling and fighting go on for a long time. I think about waking Champ, but I don't want to put myself in the middle of it. I step outside and tell Jesse they should cool it. "Champ is not going to want you drunken assholes fighting in his front yard with your fuckin' shirts off," I tell him.

"But nobody's tapped out yet," Jesse says. "I was in the marines for a couple weeks, and the only thing I learned how to do is choke somebody out."

I watch him put his skill to good use as he and the other guy choke Frank out repeatedly. Then when he is unconscious, they put the weight of their knees on his back and grind his face in the dirt. But Frank refuses to give up. He comes to screaming and cursing, fighting back, his arms and legs flailing, spittle flying from his mouth. I walk back inside.

"We must have choked him out a dozen times," Jesse would later say. "He never tapped out. You gotta hand it to him, I guess."

Gabriel and Cholly, the new Jamaican roommate, come downstairs. We stand in the kitchen while they cook. "Those guys drank a lot of alcohol, man," Cholly says.

"Those dumbass fucking crackers can't handle their booze," I say. They look at me, and I feel a flush of embarrassment. I am drunk myself, I realize. They shake their heads ruefully. I soften my tone and tell them I don't think there is anything to worry about, but they look worried. They are gentle, sweet men.

Earlier in the evening, Cholly had joined Jesse, Skip, Frank, and me while we were watching the game. We talked about women, and I passed my computer around so everyone could check Facebook and we could show each other pictures of who we were dating, or at one point had dated or wanted to date, or whatever. Skip asked Cholly if there were any light-skinned girls in Jamaica. What about Jews?

Cholly answered him dutifully, and Skip expounded on his thoughts about light-skinned black girls and Jews. Then he turned the conversation to Ronald Reagan. "Ronald Reagan is the greatest president the United States ever had," he told us, unprompted. He said it again. Then he said it a third time.

Finally, I took the bait. "What about Abraham Lincoln?" I asked him.

"Lincoln just wanted to free the slaves so he could send them back to Africa," Skip told me.

When it was Cholly's turn on the computer, no one but me looked at the pictures of his girlfriend, a black-skinned beauty. Not long after, Frank said something about "niggers." I told him to shut the fuck up and Cholly acted as if he didn't notice, but soon after that, he got up quietly and left. When I then pressed Frank about it, he mumbled and shrugged. Jesse said, "It's fucking Williston, man. Shit's different up here."

Shit is different up here. Williston doesn't feel safe to me, a flannel-clad white man who drives an SUV. But for the Jamaicans— black men who wear soccer jerseys and ride their bikes to work? The danger has entered the house.

Gabriel and Cholly take their meal upstairs, and the front door opens. Jesse clambers into the living room and falls down, his head swimming and his mouth loose. His shirt is off, and he's so scrawny I can see his ribs. I pick him up and carry him to the bathroom. He crumbles over the toilet. Kneeling in front of it, he wraps his arms around the bowl and empties the contents of his belly.

When he finishes throwing up, I tell him to sit up. I help him out of the bathroom, set him at the kitchen table, and give him a glass of water and a can of ginger ale. I tell him to drink it. He doesn't listen. He shakes his head vaguely, his slender shoulders stooped over his lean frame like a weary question mark. He leaves the drinks on the table and slides his body sideways into the living room, falling on top of his mattress and passing out.

I look at him sprawled out on the floor. He reminds me of myself when I was nineteen: clueless, impressionable, and drawn to guys who get in trouble. We don't know it at the time, but he had just broken a rib.

SONGS

I stop for breakfast at a local diner. The waitress is friendly. "You aren't working today?" she asks me. I tell her I just hired on. She says, "Well, enjoy your day off. Probably won't see you for a couple months once you start."

She is right. Work as an oil field hand will prove to be so overwhelming, the hours so all consuming, that none of the budding relationships I have been nurturing in Williston—beyond the oil field workers I am yet to meet—will have any chance to bloom.

Before moving to town, I reached out online to a local songwriter, and she invited me to join a gathering of players at the park in Watford City. I drove down one afternoon a few days after moving into the flophouse.

There was a band shell and a small grassy area with picnic tables. A group of local women sold cookies and single-serve bags of Doritos filled with ground beef, lettuce, American cheese, and diced tomato. They called them "Tacos in a Bag" and they handed them out in front of a sign that read MOPS, which, I learned, stood for Mothers of Pre-Schoolers.

There were six or seven gathered musicians. We made small talk for a few minutes and then shambled up onto the band shell. A couple of the guys were serious players. One older man wore a cap that read DAKOTA AIR and breathed through a portable oxygen tank while chain-smoking cigarettes. His face was ashen, but he had a big, warm smile. He played the Dobro and his touch was sublime. Another standout was a handsome silver-haired guitar picker dressed in jeans, cowboy boots, and a denim Harley-Davidson jacket. He had an air of easy confidence about himself and he cussed a blue streak. Equipped with a modest but effective singing voice, he sang two great songs about women leaving him.

Microphones were set up on the stage, and we took turns singing into them and backing each other up. The weather was perfect and a handful of locals listened politely.

I backed up the other players as best I could and took the lead singing a couple John Prine songs and some Woody Guthrie tunes. I felt an immediate kinship with the group I was playing with. The guitar turns strangers into friends and friends into communities. I am incredibly close to several people with whom I have barely said more than a few words beyond "What's that change at the bridge?"

I was first awakened to the power of song when I was around ten years old. My sister Kate and I sat wide-eyed and open-eared around a record player at the house of a neighbor who had a copy of *Bob Dylan's Greatest Hits*. The sound, primordial and immediate at the same time, held both of us in a weird sort of trance. As I grew older, I traced Dylan's songs back to the music of previous generations, to the murder ballads, work chants, sea shanties, and blues, songs that themselves stretch back centuries and longer, to the Middle Ages, to the dawn of Christianity and before that. On my quest for the heart of the music, I found the songs of a traveling sign painter named Woody Guthrie.

Guthrie had been born in Okemah, Oklahoma, less than five years after the former Indian Territory achieved statehood. He heard the tribal chants of the Cherokee, Chickasaw, Choctaw, Creek, and Seminole from his father and learned traditional Scots-Irish balladry from his mother. While the influence of Appalachian music in his songwriting may seem obvious to most listeners, the tribal chants would perhaps have a greater impact on his vocal delivery—a style of singing that would be copped by Ramblin' Jack Elliott, then by Bob Dylan, and then by everybody else. In addition to that, Okemah was an oil boom town, and in those pre-radio days, the men who came to work it brought with them ethnic folk music from all over the globe. Woody was the type of guy who always came in the back door, but he stood at the apex of a thousand musical currents.

Woody took the melodies of popular and ancient music, Carter Family songs, blues as sung by Lead Belly, and spirituals, and he added his own lyrics to the old tunes. Like a satirist, he wrote songs in conversation with the songs on which they were based. The double meanings deepened the music. Woody took the hillbilly standard *Jesse James* and turned it into the song *Jesus Christ*, reframing the son of man as a Robin Hood–style outlaw cut down by rich gangsters. He could be unpredictable, angry, and political. Most famously, he wrote "This Land Is Your Land," in reaction to Kate Smith's rendition of "God Bless America."

In the mid-2000s, I wrote a play about Guthrie and acted in the role of Woody. It sounds counterintuitive, but the play was the most autobiographical thing I'd ever written. By playing the role of the itinerant folk singer, I was able to investigate my own life, to explore some of my deepest preoccupations, fears, and aspirations. Woody and I had both lost family members to terrible accidents when we were eight years old, and I believe those events instilled a deep sense of injustice in both of us that would define the length of our lives in ways we would struggle to ever fully understand. While Guthrie's traveling is often given a kind of mythic gloss, the truth is he couldn't sit still. As I'd come to realize myself, later, after leaving North Dakota, the line between running toward and running away ain't always so thick, and you can only land in so many states before you have to start asking yourself what is chasing you.

Guthrie wrote plays and novels, and he composed thousands of songs. He wrote songs about working people: fruit pickers, immigrants and migrants, farmers, truck drivers, miners, and sailors. He was drawn to the down-and-out and, as his daughter Nora told me, "This wasn't a political decision for him. He just felt more comfortable around people like that." He was one of them, I recognized, and that didn't change when he played a show on Broadway or hung out with a pop star. I wondered if, like me, he ever wished it would change. He wrote all kinds of songs: spiritual songs, children's songs, righteous protest music, filthy sex songs, sweet love songs, battle cries, and unholy chants. He told stories in his songs, documenting the lives of working people throughout the whole middle of the twentieth century.

Performing the role of Woody in my play set me on the path toward playing guitar professionally. I'd written songs and fooled around for fun, but knowing I'd be performing for hundreds of people raised the stakes and forced me to work on my technique. As I gained confidence through that, I started booking more of my own paying gigs.

I'd spent my first weeks in Williston looking for places to play guitar, but it was an almost comical quest, dropping into bar after bar advertising LIVE MUSIC only to be told Tuesday was karaoke night. One of those first nights, I saw Lana Del Ray on TV. We'd performed at some of the same small New York clubs before she became famous, and we had developed a casual friendship. I turned to a couple guys sitting next to me at the bar. "Hey, I know her," I said. It was exciting to see her on TV. The first time I met her, I'd come off stage after playing a short set, and I saw her across the room, clapping brightly and smiling. She met my eyes and mouthed the words "That was good!" and it all but knocked the wind out of me—not only was she beautiful, but she emanated openhearted kindness. And now here she was: one of the world's biggest stars, and I felt bound to her by a camaraderie I think only music can provide. It didn't matter that I was sleeping in my Chevy and she was floating by a pool in Malibu or wherever. I felt connected to her. The guys at the bar, however, looked at me with passive expressions of such bored disbelief that suddenly I felt like maybe I was lying. Embarrassed, I turned back to the TV.

The reality of Williston at that time was that, beyond a guy in the back room of J Dub's playing an Ovation guitar along to a drum machine, I couldn't find any live music in the city. It might sound minor when talking about a municipality wrestling with the problems Williston was dealing with at the time, but I don't think so.

In Williston, the lack of music felt dangerous. There was a gaping hole in the heart of Williston where music should have been. The place had a case of laryngitis. Boomers had no voice, not just politically, which is obvious, but to sing along with, which is possibly more important.

The evening I spent playing songs in Watford City was a balm to my lonesome soul. I wasn't alone when I was trading tunes with other pickers. The act of playing put me squarely back into a community of musicians. I didn't know the players yet, not personally, but I knew where I fit among them. And I got a thrill from being onstage, like I always do, stomping out rhythm and melody on a six string, hollering my words and singing folk songs, feeling the timeworn melodies and rhythms move through me. It was one of the very few moments of easy joy I would experience that whole summer.

After the gig, we chatted and exchanged numbers. I promised to follow up with them but a few weeks later I'd land a job, and I'd never see any of them again. Work would keep me too busy. I'd be unable to maintain any of those first connections I made, but I'd think of those players often.

By joining the ranks of oil men, I would not only be giving up my connection to several individual people but also be losing any chance I had of really connecting with the town of Williston outside of how other oil patch workers experienced it—as a place to eat, sleep, and shit. A bookstore clerk was kind enough to hand me a flyer for a local theatrical production, but I'd be too tired to attend. The young women at the local coffee shop would continue to smile at me warily, but I'd never be in any situation to ask one of them out to dinner. I'd never make it to the county fair, to a Native American powwow, or to see a local rodeo. I'd be working.

Locals resented boomers' lack of involvement in the community, and boomers routinely referred to Williston as a "shit hole where there's nothing to do." Both were right and wrong at the same time. Crude has a way of flooding the culture, disappearing connections in charcoal plumes of smoke, making everything greasy and black.

When I pay my bill, the waitress tells me she is from Idaho. "Going back in two weeks," she says. "That's why I'm so happy. Put in two years here. Living in a trailer, all that. Two years. It's like a prison sentence. I'm going back. Good luck."

THE SAFETY MAN

The Safety Man is a round man with a large square head, like a gift on top of a beach ball, and small hands. He breathes loudly through his nose in long, steady inhales and exhales as he leads me to a cold paneled basement room, shows me a small stack of VHS tapes and a twenty-inch color TV, and leaves.

I spend several days in the basement of the dispatch building watching old tapes. In one, an actor playing Death, who is called "Big D," conspires with an actor playing Hydrogen Sulfide, or H2S—an odorless, colorless chemical—to kill workers. The video uses POV shots to show H2S chasing down workers like the shark in *Jaws*. In another video, a sling salesmen talks about safely rigging synthetic slings. The salesman has a thick mustache and styled hair. He wears a slick western jacket and a bolo tie. He says, "Any limp-wristed office wimp can take this product out and rig a hundred-eighty-thousand-pound load with it." I have no idea what he's talking about.

I get sent to OSHA 10, ten hours of safety training spread over two days as mandated by the Occupational Safety and Health Administration. My first instructor is a terminally genial pastor. He shows us macabre industrial videos where workers fall off ladders and get their hands stuck in machines. They are hit in the head with steel beams. One worker is set on fire. I learn that safety comes first.

I start thinking about Buck, the guy from Missoula who hit town the same day I did. We'd been in touch during our job searches, texting each other encouragement. Weeks earlier, when I was at my lowest, feeling fearful nobody would hire me on, Missoula Buck landed work as a roustabout. "There's opportunities out there," he said. "You just got to grab them." It gave me hope at a critical moment. If he could do it, so could I.

I send him a message telling him I got a job. He responds with a torrent of texts. "You've got real determination," says Missoula Buck. "You are most certainly going to achieve your goals as long as you do what is in your heart that you know is right."

The Safety Man summons me to his office. He gives me a green hard hat, two pairs of safety glasses (one clear and one shaded), a bag of gloves, and a single fire-resistant, or FR, jumpsuit. We climb into his SUV and drive north toward the job site.

Location is a wide, flat, dusty swath of land in the middle of the prairie. It looks like the moon landing. We sit in the SUV with the AC pumping and stare out the window at what I would come to know as Sidewinder Canebrake Rig 103—the first drilling rig I will ever break down and the first rig I will have a hand in putting back together. At the time, I have no idea what I am looking at. I have no way to describe it but as a big, metal *thing*. It is surrounded by trucks and cranes and men in hard hats and jumpsuits. I feel like I am staring at a dangerous book written in a different language.

The Safety Man points at a worker walking behind a giant steel platform. "Look at that fucking idiot," he says. "Never stand under or in front of anything that can crush you." He points at a worker who is leaving the company, the swamper I am replacing. "We tried to train him up," he says, "but he couldn't drive a truck. Can't back up, I guess."

I am introduced to no one. We never leave his SUV. Back at the yard at Diamondback, the Safety Man drops me off. He tells me to report to dispatch the next day at 6:00 a.m. Before leaving, I try to ask him a question, but he cuts me off. "Good luck," he says.

That will be the extent of my safety training.

When I get back to the flop, I grill sausages in the backyard. "So are you psyched?" Champ asks me, standing by the grill. At first, I'm not sure what he means. "Everybody comes out here to get a job in the oil field," he says, "but most of them end up working at Hardee's or some shit. You did it." I am excited, I realize. And nervous.

I eat my dinner at the picnic table and share my beers with Champ and Rickie as they climb into the hot tub. "If you want to get in," Champ tells me, "you can get in with us."

"No, thanks," I say.

"Seriously," he tells me. "I usually don't let anybody get in here with us, 'cause I don't want everybody to think it's open season or some shit. But if you want to, you can get in."

I beg off. The idea of climbing into the hot tub with Champ and Rickie is weird to me. But it also feels weird to refuse and, you know, hot tubs are nice. So I grab my swim trunks and hop in. I sit there a moment in the hot water and bubbles with Champ and Rickie. We all kind of look at each other. "You know," says Rickie, "there's this fantasy Champ and I have been having . . ."

"Oh, be quiet," says Champ. I nearly blow beer out my nose. The whole scene makes me want to giggle until I have an aneurysm, but the hot tub feels good on my back. We sit in a kind of awkward, contented silence until Champ asks me how I feel about the job.

"Good," I say. "I think."

"You'll be in the oil patch," he tells me with a certain finality in his voice. I realize I don't have any idea what that means.

ROOSEVELT AND ME

In 1884, the grieving twenty-five-year-old New York State assemblyman Theodore Roosevelt—after losing his mother to typhoid fever and his wife to kidney failure on the same day that his first daughter was born—marked an X in his journal and wrote, "The light has gone out of my life." He then headed west to North Dakota.

A scrawny young man with a high-pitched, scratchy voice and a nervous disposition, the foppish dude from New York had, the previous year, ridden into Little Missouri the day after the completion of the Northern Pacific Railroad corridor, on an expedition to hunt buffalo. By horse, he traveled deep into the Little Missouri Valley, scouring the badlands for bison throughout a weeks-long torrential downpour.

An easterner, like me, Roosevelt was taken with the rugged beauty of the land and he wrote of it vividly:

> The grassy, scantily wooded bottoms through which the winding river flows are bounded by bare, jagged buttes; their fantastic shapes and sharp, steep edges throw the most curious shadows, under the cloudless, glaring sky.

On this first trip to North Dakota, after turning it over in his mind for barely a week, Roosevelt decided to invest a full third of his substantial fortune into the cattle business. He handed a check for $14,000 to a ranch hand he'd just met. Within a couple days, he'd killed his first buffalo, dancing around the felled beast "like an Indian war-chief, whooping and shrieking" in excitement at his victory.

As he would later recall, "It was here that the romance of my life began."

Roosevelt—a victim of debilitating asthma attacks and prone to persistent stomach problems and diarrhea—had been advised by doctors to lead a quiet sedentary life or risk an early death. He decidedly ignored this advice when he returned to the hard soil of North Dakota, substituting the drawing rooms and chatter of New York politicos for the dry prairie wind and the songs of bluebirds, whip-poor-will, mourning doves, and meadowlark.

The New Yorker cut a strange figure in the cattle settlements of western North Dakota. In a perfectly tailored cowboy costume of fringed and beaded buckskin tunic, broad sombrero hat, horse hide chaps, cowhide boots, and silver spurs, the "dude Rosenfelder" was at first relentlessly mocked for his reliance on spectacles.

But the big nerd earned the respect of the iron-willed men of the West. Roosevelt oversaw a herd of a thousand cattle, riding as much as a hundred miles a day on his trusty steed Manitou. He wrote *Hunting Trips of a Ranchman*, traversed the prairie up and down the Missouri River, hunted and killed literally thousands of small and big game, slept in his saddle during roundup, and punched out a bully in a barroom. He even captured a trio of boat thieves, forcing them at gunpoint on an eight-day journey to the local authorities in the town of Dickinson, all the while reading and completing Tolstoy's *Anna Karenina*.

ROOSEVELT WOULD LOSE a substantial amount of his fortune in North Dakota, perhaps more than half of the $80,000 he would eventually invest. The cattle boom that started with a wild scheme by a mysterious Spaniard, the Marquis de Morès (the town of Medora is named for his wife), ended in a bust after the Winter of Blue Snow—a series of snowfalls and blizzards starting in November 1886 that left the region's cattle country completely decimated.

But Roosevelt's experience in North Dakota would transform him. The squeaky-voiced, four-eyed dork with the struggling breath and persistent case of the shits, the kid so radically skinny his hips barely kept his britches up, metamorphosed in the open-aired country of North Dakota into the Rough Rider of popular imagination. Roosevelt would return to New York husky, lusty, and brown and tough as a hickory knot.

Teddy Roosevelt was the Indiana Jones president. A man of an acute intellect who was also a man of action. He went to North Dakota, like me, and he became a man who would be president.

I have no illusions about going into politics, but I identify with Roosevelt's transformation. I was the skinny kid in middle school who wore an army jacket to class every day—my father's army jacket. I had been hungry to find something in my old man that I could admire, and when I came across, tucked away in a dusty closet, an olive drab coat with sergeant's stripes down the sleeves, I put it on. My father never made much of his military service, but I was obsessed with it.

Dad was forty-five years old when I arrived in the world. A decade and a half older than most of my peers' parents, he'd been born one year after the stock market crash of 1929. As a child of the Depression, Dad was raised primarily by his mother's family, an unruly clan of Scots-Irish called Boyland, in the Ridge and Valley Appalachian town of Cumberland, Maryland. Like his younger brother and older uncles, he took up boxing as a kid. As a young man, he served in the Korean War in the 187th Airborne Regimental Team—nicknamed The Rakkasans by the Japanese, which literally means "falling umbrella men." He started the war as a rifleman second scout and was promoted to machine gunner when the first machine gunner was killed with a burp gun blast that laid open his forehead. Dad rarely talked about it, but to say he "probably shot a few trees" and reminisce about the whores he bought in Japan for a dollar each, but his service loomed huge in my imagination.

He had volunteered. That was important to me, and growing up, this felt like a potential connector between my father and me, a hinge we could both swing on. My dad was US Army Airborne. He had literally leaped out of airplanes into machine gun fire. I lusted for that kind of action, and I thought my aspiration bound us. I thought that if I went through what he went through I would gain not just insight into his life but respect for him, too.

My family is full of military men and adventurers. My father's uncles were true WWII heroes. One of them, Joe Boyland, was captured by the Japanese and pressed into service as a truck driver in the Philippines. He escaped captivity and joined a guerrilla cell in the northern mountains of Negros Island, organizing attacks on the truck routes he'd been forced to drive. Joe fought in the insurgency for nearly a year before he was evacuated and returned to the States. His brother William, whom I'd be lucky to know, had joined the navy. Bill was torpedoed at least once during WWII. He survived and, after the war, signed up with the Military Sea Transport Service, where he eventually captained his own ship, transporting arms and soldiers throughout both Korea and Vietnam. Upon retirement, Bill became a pilot on the Panama Canal. He passed away in 2011, but he remains the greatest storyteller I have ever met in my life. Captain Bill Boyland had swung around this little blue planet so many times that to him life was a small thing, the earth the size of a grapefruit.

What had I done? By the time I finished high school, I knew I wasn't going to join the military, but it wasn't an easy decision. I spent a great deal of time wrestling with it, scheduling phone calls with recruiters and then canceling them, poring over army brochures, then stuffing them into the trash. I felt very alone. There seemed to be no one I could talk to about it. I'd fallen in love with theater. Who on earth is torn between boot camp and acting school? I was.

But I decided that if I joined the army that would be it: I'd be an army guy. Whereas if I pursued life as an actor, I would be entering a world where the potential for adventure was as vast as words written on a page. Then that dried up. I wasn't getting any acting gigs. I was pushing paper as an administrative assistant, living a post-adolescent life in a post-adolescent Brooklyn full of office workers costumed as lumberjacks. Like a pre–North Dakota "Rosenfelder," my world felt frivolous. It revolved around leisure. I wanted out. I was too old to join the military, but I could become an oil field hand.

STANDING IN THE BACKYARD of the flophouse, a day shy of starting work as a field hand, I know somewhere deep inside myself that I want to get my ass kicked. I want to be beaten and pummeled and knocked the fuck down. I want to find out if I have it in me to get back up. Because if this breaks me, then I'll know what breaks me. I'll know where that line is. I will know I tried and I lost, and I will know where I lost and how hard I tried. And if I power through, if I come to a place where I feel to myself that I have proven myself, then I will know that. It will draw for me a clear perimeter in what still feels like a boy's life, a life without boundaries, without coherent narrative, without meaning. It will grow me up. I want to grow the fuck up.

BOOK TWO

THE PATCH

—

I was born working and I worked my way up by hard work.
I ain't never got nowhere yet but I got there by hard work:
Work of the hardest kind.
I been down and I been out
And I've been busted, disgusted and couldn't be trusted.
I worked my way up and I worked my way down.
I've been drunk and I've been sober. I've had hard times
   and I got hijacked
And been robbed for cash and robbed for credit.
Worked my way into jail and outta jail
And I woke up alotta mornings and I didn't even know
   where I was at.

—WOODY GUTHRIE

DAY ONE

I arrive in the bleary morning at Diamondback Trucking's dispatch office wearing work boots, jeans, and a T-shirt, carrying my green hard hat and my new jumpsuit. Dispatchers sit in mismatched office chairs behind a glass wall taking phone calls, tapping at greasy computers, blowing on hot coffees, and grunting back and forth with truck drivers. The yawning truckers line up at one of two windows to scratch their bellies and pick up photocopied map routes. Scruffy-faced, dirt-stained pipe yard workers bust in and out of the back door to punch their time cards. They enter and exit loudly, in midsentence, already cussing at 5:45 a.m. In a show of their higher status, the oil field hands enter dispatch through a doorway next to the windows and sit behind the glass wall with the dispatchers. They shoot the shit, chew dip, complain, sip Red Bull or coffee, razz each other—killing time before leaving for the job site. I stand in the hallway and try to take it all in. I don't know what to do with my hands.

"We got a *greenhorn* starting today." A pipe yard hand laughs and looks me over. The word is cowboy slang for the new dude, untrained and inexperienced as a young steer. It was literalized in the oil field where newbies are given a green hard hat while experienced guys—those who have worked six months or longer—wear white ones. I meet the eyes of the pipe yard worker and do my best imitation of an easy smile.

The door to the yard opens, and a giant kid seems to fold himself in through the doorway and then unfold himself in the hall—he's so tall. He struts down the hall wearing a hangdog expression under dark sunglasses and blue jeans, a black sleeveless T-shirt that says in big block letters: I DON'T CARE HOW YOU DO IT IN TEXAS, OKLAHOMA, MONTANA OR IDAHO. THIS IS NORTH FUCKING DAKOTA.

"This is Michael," a dispatcher deadpans to the big guy, introducing me. "He's starting today. Take him under your wing. Show him everything you know, okay."

The big kid's hangdog look is swallowed by a toothy Eddie Haskell smile, "Well, I don't think that'll be too much then."

I follow him outside into the dew-wet predawn dark of morning. We climb into a white van filled with old-timers and young guys with short cropped haircuts, dirty shirts, and sun-stained faces. They sit staring straight ahead, exchanging occasional grunts and mumbles with their hands in their laps, like well-behaved schoolkids going on a field trip. No one says a word to me as the door closes and the van rolls over the gravel of the yard and swings out onto the highway.

The giant kid folds himself up in his seat. His head drops back on a loose neck and, within a minute, he falls into an openmouthed, lightly snoring sleep. I am anxious about the day ahead, more wired than tired, but I mimic the kid and play possum, leaning my head against the window with my eyes closed.

The van stops at a convenience store about thirty miles east in the town of Ray, and everyone piles out to grab breakfast. I buy a watery coffee. The giant kid scoops up a personal pan pizza from under a row of heat lamps. In the van, he demolishes the pizza in three crunchy bites and drops right back into drooling oblivion.

When we arrive at the work site, the big guy wakes and yawns and stretches. We hop out of the van and onto the dirt. The air has a snap to it, the sun barely reaching up over the plains. We stand behind the van and take off our jeans and T-shirts. I feel small compared not just to this giant kid but also to the scale of the earth and sky. My skin pale, my arms thin, I'm a speck on a square foot of dirt out amongst an awesome glacier-cut sprawl of land and time. We step into the heavy cotton fabric of our fire-resistant jumpsuits and zip them up, nothing underneath but boxer shorts and the pink of our skin.

I ask the kid about the job.

"How old are you?" he says.

"Thirty-six," I tell him and feel a mild panic.

"Well, you still got all your fingers," he shrugs. "You'll be fine."

The kid, like Mark Twain's hero, is called Huck. Huck is six foot seven and aw-shucks friendly, with an open, rubbery face and a small half-moon scar under his left eye.

"I just came back from Kalispell, Montana," he brags. "Had my twenty-first birthday there, bar hopping with friends. I was kicked outta the first bar for wearing my hat backwards. I downed so many Crown and Cokes, I think we were kicked outta a second bar, but I don't even remember."

He got himself a tattoo for his birthday, and he pulls his pant leg up to show it to me. A busty pinup girl swings from the wrecking ball of a crane. "It's like a painting," he says. Then he adds, "She's pretty hot."

"What's it mean?" asks a truck driver.

"You know," Huck responds, a grin swallowing his face, "oil field trash!"

Huck and I are swampers. A swamper is a truck driver's assistant. Trucks in the oil patch don't carry passengers, so the passenger seat is called the swamp seat. A swamper is the feet on the ground for the truck's operator, backing the driver up in tight situations and rigging any material the truck needs to haul. A good swamper is invaluable to the work of a good operator. "A swamper is a trucker's bitch pretty much," Huck says.

On my first day, Huck and I are swamping for the crane. Crane swampers are also called riggers, but the terms are more or less interchangeable, and the job is essentially the same.

On this day, we are rigging down Sidewinder Canebrake 103, one of the most technologically advanced drilling mechanisms in the world. Oil rigs are called rigs because they are rigged up and rigged down. Trucks are also called rigs. Rigging is a verb as well as a noun. As a verb, it describes the action of putting the pieces of an oil rig together (rigging up) or taking them apart (rigging down). As a noun, rigging refers to any piece of material that is rigged up or down as well as the tools used to do it. Every piece of rigging is rigged by rigging. "Got it, dipshit?"

Huck and I climb up onto the crane, and the operator climbs into its cab. Once Huck shows me how to arrange the chains and slings on the crane's line, we then hop down and walk over to a series of large, steel-reinforced, wooden-planked mats roughly eight feet wide and forty feet long.

When an area is chosen for drilling, the work site, often referred to as the oil lease or more simply as "location," is leveled off and filled with chips of scoria, a cindery, vesicular basaltic lava. Scoria is the same type of rock that makes up the rust-colored crags of North Dakota's badlands. Once the scoria is spread, because oil rigs weigh thousands of tons and need to be level to operate, large wooden mats are laid across location, and the rig is built on top of the mats. On my first day of work, rigging down Sidewinder, we are moving the mats onto haul trucks that will take them to the next location, where the rig will be put back together.

The four chains dangling from the crane's headache ball drop onto the center of the mat in a rattling thud. Huck and I grab the hooks at the end of the chains and slide them into pockets in each of the mat's four corners. Then we stand back. The crane flies the mat into the air and swings it across location. We follow behind, watching it soar through the window of blue sky. When the crane operator lands the mat on the bed of a waiting tractor trailer, we unhook it from the chains, then walk back to pick more mats.

At one point, a stack of mats is suspended in the air above the bed of a trailer. Huck is swinging them clockwise to line them up with the trailer bed, so the crane operator can drop them. But he is having trouble. I hesitate a moment. I'm not sure what to do. "You can put your motherfucking hands on it! It's not going to motherfucking bite you!" the crane operator yells at me from his cab. I quit hesitating. I move in close and put my hands on the mats, trying to gauge where to stand so as not to put myself in danger. I push. We swing the mats into position, and the operator lands them on the trailer's bed.

"Some guys are dumb fuck assholes," Huck tells me, kneeling to insert a hook into the pocket of another mat. "You should know that." He shrugs. "Don't let it get to you."

The hours tick by. There is a simple joy in the completion of a repetitive physical task. The sun, hot and close, moves across the sky. The oil rig is half apart, strewn across the mud like a distracted child's unfinished Lego skyscraper. A light breeze floats through us.

Abruptly, the crane operator puts his hard hat on and leaves the cab. Huck stops what he is doing and follows him to a pickup truck. I follow Huck and sit in the back of the cab. The crane operator pauses to look out the window at the rig. "That motherfucking rig is the dumbest fucking thing I've ever seen," he says. Then silently, he pulls the top off a cold can of Chunky soup. Huck follows suit with his own can. I unpack the sandwich I brought and listen to the sound of spoons scraping against tin as quietly we eat. When he finishes, the operator puts his seat back, closes his eyes, and falls immediately asleep. Huck leans his big head against the passenger window and starts snoring. I close my eyes, lean my head back, and doze. It feels like a mere moment later that the radio crackles and a voice summons us back to work. I wipe the drool from my mouth, and without talking, we all get out of the vehicle and get back to it.

Even after lunch, my stomach feels empty and alert as if, instead of a ham sandwich, I had swallowed an exclamation point. My head is glowing in a way. My eyes are clear and my thoughts keen as a hunter's. I move through the swirl of action—the wind, the dirt, and the machines—and breathe it all in and out. I know this is fear, but with it has come clarity. A breeze tickles the hair on the back of my neck. Sweat evaporates off my skin. In the window of a pickup truck, I catch the image of a field hand: FR jumpsuit, hard hat, and shades. My own reflection. I look exactly like all the men around me. I've already become one of them in a way. It feels like opportunity, a chance at some kind of transformation.

"Is this your first day?" a truck driver stares at me through dark glasses.

"Yeah," I tell him.

"You scared?" A grin slaps across his stubbled face. "You look scared."

"I'm not scared," I lie. "But I'm being careful." He keeps grinning and keeps staring. So, I walk away from him, back across the scoria to the crane. Fat clouds move steadily across the horizon silent and unalarmed. I wonder how they do it.

Abruptly, the crane stops again. "Grab your shit from my truck and put it in the van," the operator tells me. "We're leaving, but you're staying here."

"The only way to figure this stuff out is get your hands on it," Huck tells me before hopping in the truck with the operator and leaving location. I take a deep breath and march across the scoria into the roar of the machines.

HUCK GETS BOUNCED

There was something about Huck that I was drawn to immediately. He was larger than life. Literally, for at six foot seven, Huck loomed over everyone around him. He vacillated between timidly trying to mask his big size, shrinking in on himself—his neck leaning forward and shoulders dropping—and proudly bursting out of his own seams—sauntering, chest puffed out and arms thrown back. He'd been a fat kid, he told me, and because of this, he sometimes exhibited a fat kid's paralyzing insecurity. But when he settled into his skin, Huck was big. A giant walking in a sloping gait, his arms so long that his knuckles nearly scraped the earth.

In the time I'd know him, Huck would get in nearly a dozen fistfights, he'd be arrested and thrown in jail at least three times, he'd get blind drunk and spit on a police officer. He'd have his head bashed in behind a local bar, he'd steal a gin truck for joyriding, and he'd operate a crane on location without a license. Through all of it Huck never lost that shining thing. It may seem outrageous, but in retrospect I can only call it innocence. It was the sun at the center of his soul. His fists and drinks and reckless driving were just the planets that spun around it.

Huck loved big trucks with the unadulterated glee of a seven-year-old. He loved the dirt: the patch was a big ole sandbox to him. The more grease smeared across his nose, the more diesel soaking through his FRs, the more gunk deep down under his fingernails, all the better. He loved fights. He would deny this. "I don't know why these fuckers are always starting shit with me," he'd say mournfully, but he loved them. It was roughhousing—taking a tumble in the dirt with a drunk behind a bar or throwing down bare-knuckle against a couple buddies after one-too-many bottles of booze. I think it made him feel closer to the men around him.

Not that he didn't regret it afterward. Huck was forever getting into fights and then feeling bad about it. Huck feeling bad was hilarious. He would all but set up a chair in the corner and sit on it with a dunce cap and a frown painted on his face.

I've never known anyone who could swing from such gleefully high spirits to absolute down-in-the-dumps melancholia in a snap of the fingers like Huck. It amused the hell out of me. He would be doubled over in guffaws midstory about slamming a dude's head against a tailgate when a sudden stricken look would cross his face. "I hope I didn't hurt him too bad," he'd say, as if the thought had only just occurred to him. His rubbery visage would fall in an avalanche of despair that utterly transformed him from a bright, toothy fox-with-an-egg-in-its-mouth to the hangdog countenance of a chastised basset hound.

As we'd grow closer over the following months, I'd see more of this joy, this innocence. It was never far from the surface of his skin, and I gleaned it right away in the way he told his stories, in the way he parsed his brags, in the way he mouthed off half-proud, half-embarrassed, in the way he laughed at himself and the situations he found himself in.

"We were at Heartbreakers looking at titties," Huck tells me. "We were kinda broke and Smash pays for our drinks with quarters, and the bartender comments on it. He says 'You gotta be kidding me! Where's my fucking tip?' So, I said, 'You're an asshole, and you're not getting a tip.' I was pretty straight to the point and polite about it."

"Uh-huh," I say.

"Well then, here comes this bouncer, man. He's big. Tall as me, and he's got a Mohawk. He says to the bartender, 'You want me to bounce him?' Like he's Patrick fucking Swayze, you know? The bartender is like, to me 'You can either apologize or finish your drink and leave.' So, what do I do? Right hand lifts the beer to my mouth, left hand gives him the bird. Well, here comes homeboy. Mohawk slaps the drink out of my hand, grabs me by the throat, and starts pulling me down. But he can't get me down 'cause he's a fucking uncoordinated bitch. But then his two buddies show up. So one hooks me in this arm. One hooks me in that arm, and *boom!* they throw my ass on the ground. And what does Mohawk do?"

"Lemme guess."

"He fucking jump punches me! And my head is on the ground. He puts a dent in my forehead, for fuck's sake! He's like *boom* hits me. And I'm like 'holy fuck.' I'm dazed. I get up and I'm looking for him, and he ran, dude. He left! And he called the cops on top of it."

"Wow."

"I saw him the next day at the library, though," Huck says. "I was like, 'What the fuck, forehead puncher? You're kind of chicken shit, don't you think, running off?' He said, 'I'll show you chicken shit. Go over to the truck.' And then we went over to his truck and talked it out. It was like daytime, you know. I guess it would have been dumb to fight him at the library. He had some of my paper-work."

"He had what?"

"My alcohol evaluation paperwork. I had it in my pocket that night at the bar and it fell out when he punched me. Is that weird?"

"Wait, what paperwork?"

"For the DUI I got last Christmas. I needed it for the court system, to get my license back. I spent like two hundred dollars on that paperwork."

"Yeah?"

"He was like, 'Is this yours?' and I was like 'Yes, it is. Thank you.'"

COMINGS AND GOINGS

Frank sits on the living room couch with a fitted, flat-rimmed Yankees cap cocked sideways over his bald pate. He wears a large T-shirt with images of Tupac and Biggie on it.

"Didn't they hate each other?" I ask him.

He shrugs. "I got a second job at a hotel," he says. "They're giving me housing for two hundred a month."

"That's great," I say.

"I might even split the room with somebody," he says, "then I'll only be paying a hundred a month."

"Right," I say. He has a couple suitcases at his feet. "Good luck," I wish him, and he is gone. Before I can blink, Jesse's uncle is there in his place.

Jesse's uncle Ernie had a good-paying job, but he got drunk and wrecked the company truck, and he's been sleeping in the park ever since. His face is a slab of meat. Beady eyes set deep and dark against his skull, he has bad skin and a kind of formless body. The first time I lay eyes on him, he looks like a little kid, sitting cross-legged on the floor staring at the TV, drinking booze and soda out of a large plastic cup.

. . .

THAT NIGHT, my sleep is interrupted by something guttural, gastrointestinal, and horrendous. A rumble turned into a whistle turned into the sound of a man choking to death. I wake up. It's Uncle Ernie. He is sprawled across the couch with his feet inches from my head. His mouth is open. From it is coming the most awful sound I have ever heard in my life. He is snoring.

I bury my head under a pillow. I plug my ears. I get up and get a drink of water to settle my nerves. I try to meditate. I think about killing him. None of it does any good. I lie in bed, my body heavy with exhaustion, my mind reeling with anxiety. Precious minutes of rest pass by. I gather up my bedding and my pillow and, in my slippers and boxer shorts, I walk out onto the street, climb into the back of my SUV, and fall asleep.

I wake to the dry, painted heat of North Dakota's sun. The back window of the Blazer has popped open, and Champ stands there looking in at me—his bright orange hair a sun-filled halo.

"Sorry, dude," he says. "I'll make him get some of those nose strips or something."

DAY THREE

In the mud, I am struggling under the weight of a large wooden mat. The muck foams up over the foot of my boot, entrapping it, and making a perverse sucking sound every time I pull it out. Every step is a chore. Beads of sweat form across my forehead and trickle down the back of my neck. I use my arm strength and all of my body weight to swing the mat in front of me, and I flop it down in front of the crane's outrigger. It hits the ground with a hard slap.

"Almost too much work for a little feller!" the crane operator yells.

"Almost," I agree, catching my breath with my hands on my knees. I stand upright, "But not quite."

The 120-foot-tall lattice crane is good for making heavy picks like generators and mud tanks, but it is not easy to move. Because the ground we are on is so soft, the outriggers—the feet that keep the crane level—have to sit on heavy wooden mats, roughly four-by-four. Every time the crane moves, the outriggers are contracted, and I have to load the mats—easily a hundred pounds each, wet and muddy—up onto the body of the crane. I drop into a squat and work my hands underneath to hoist them out of the suck. I feel my muscles straining. Soon my arms are like jelly.

I trudge alongside the moving machine. The wheels spin and kick out slop, searching for purchase. I am covered in slick red muck from boot to hard hat. The mud is a combination of crushed scoria, dirt, and an oil-based drilling fluid called invert. Invert smells like diesel and looks like chocolate mousse. It is a type of "mixing mud," a special brew of water and petrochemicals that lubricate drill pipe, carry rock cuttings to the surface, cool the drilling apparatus, and balance what is called downhole pressure on a drilling rig. There's been a spill on location; the stuff is everywhere and nobody cares.

I stomp through it. The fire-resistant jumpsuit the Safety Man issued me is too big. I have to roll up the legs or trip over them; mud is getting captured in the rolls. I would guess I am carrying twenty extra pounds with me.

The crane gets stuck and can't get out. The rear driver's side wheel disappears into the muck, and the other wheels spin uselessly. A Peterbilt gin truck arrives. The driver takes a look at the situation. "Fecal matter!" he shouts. He ties his truck off to the crane like a tow truck, gets back in the driver's seat, and starts to pull forward. The crane is slow to budge. The gin truck driver leans into it. The front wheels of the truck raise into the air a few inches, then a foot, then two feet. The Peterbilt is snorting, pouring diesel from its stacks and popping a huge wheelie just as the crane begins to lurch forward. They ride, linked in metallic unison, to the other side of location. The crane gets in position for its next pick.

I take my hard hat off and wipe my brow with the back of my arm. I put the hat back on, and step by step I work my way across the mud, through the smell of diesel and the drone of machines, and over to the crane. I pull the mats off the body of the machine and slap them back into the slop. I then use the controls on the rear outrigger to extend each leg and drop each foot onto a mat.

We repeat this process every time the crane moves. Once in position, I follow the crane's chains, hook onto pieces, and fly them onto waiting trucks. We finish the picks, then rig the crane down, and move again. Then rig up, make some picks, rig down, and move again.

My body is hit with sudden bright waves of fatigue. I want nothing more than to sit in the mud and surrender to it. We rig up the crane, I toss the mats and make some picks. We rig down and move again. A ripple of exhaustion trembles through my body. I hit a wall. Stubbornly I move through it. I hit another wall and break through that. I hit another wall and barely. Just barely. I manage to keep going. I suck in my breath and keep pushing, sloppy step by sloppy step.

THE TOOL PUSHER stands in the middle of location and spits in the dirt. If he were running a construction crew, he'd be called foreman; but in the oil field there is great power in the literalism of his title: a tool pusher pushes tools into the hands of the roughnecks who work for him. He's a roughneck himself—a nickname for oil field workers that dates back to the Texas boom of the early 1900s, it literally means "uncouth person." Because drilling rigs operate twenty-four hours, drilling operations are performed by two roughneck crews, each working twelve-hour shifts. Crews are made up of drillers, derrick hands, motor hands, and floor hands. The tool pusher's job is to boss them around.

This tool pusher catches the eye of the truck pusher and waves him over. Truck pushers like tool pushers are foremen of sorts, but they push trucks. The truck pusher is my boss. He tells the truck, crane operators, and swampers what to do and when to do it.

"See that guy?" says the tool pusher to the truck pusher.

"Yeah," says the truck pusher. They are standing on the ground, the tool pusher pointing up at a roughneck, harnessed onto the derrick, swinging a sledge.

"He's one of the few," the tool pusher says.

"Oh yeah?" responds the truck pusher, taking a long look at the toiling roughneck.

"He's one of the few," repeats the tool pusher wistfully.

The truck pusher on the rig down of Sidewinder Canebrake 103 is Bob Olhouser, and he is a sight to behold. At sixty-three years old, Olhouser has a crick in his neck and a hitch in his gait. He marches through the dust patch like a damn field general, barking orders into his radio, erupting into a comically high-pitched giggle, and putting his hands on everything: no dangling chain or sedentary sledgehammer is safe from the old man's thorny grasp. Olhouser loves action, loathes the idling truck and the lazy hand. He simply can't stand stasis. So he hollers, cajoles, and conducts all movement forward. In the center of location with a radio attached to the front of his jumpsuit, pulling his top zipper down, exposing curly gray chest hair and pale skin beneath a leathery, weather-beaten face and neck, Olhouser stomps in a circle and the patch swirls around him. His square head, two-dollar haircut, and flinty-eyed stare at its center—an ornery wheel inside a wheel. It was said Olhouser believed God made humans wrong: if only we didn't have to eat and sleep, Olhouser opined, men could just work.

He likes the new guys. Like a father learning from his child, Olhouser sees the rigs anew through the eyes of the greenhorns. More than once, on those first bright, bitter days, he asks me if I like the work. "I love it," I say, and he unpeels a wild laugh, his body wiggling like he might burst out of his skin. His voice can get as

high-pitched as his laugh, too. Working for Olhouser sounds sometimes like taking orders from General Mickey Mouse.

I have no idea if I love the work. It is my immediate response, as
reflexive as a knee jerk under a doctor's hammer. Upon reflection,
I see how defensive the impulse is. If you are getting your ass kicked,
you can at least tell the sons of bitches that you love to fight.

"You ain't pooped out yet?" Olhouser asks me, grinning.

"I am plumb fucking tired," I tell him, gasping for air and sum-
moning all my bravado, "but I ain't pooped out yet."

He cackles again, gives me something heavy to carry, and stomps
off into the sunny racket.

THE OIL RIG'S derrick has been laid down and separated into two
huge pieces. One half is laid atop a haul truck's trailer. It is wider and
longer than the trailer itself, and I am told to help the driver secure it.
The driver is a little guy with white hair and a handlebar mustache.
"Forgive me if I don't remember your name," he says to me by way of
introduction. It is the most delicate sentence I have heard in days.

Handlebar shows me how to use the binders, also called boomers
or chain shorteners, to tighten the chains we use to secure the derrick to the trailer. A boomer is a metal tube with chained hooks on
each end. "They're like women," Handlebar confides. "Good for some
things. But not many." To use them, you extend the binder as much
as possible and then hook it into the ends of a chain, then crank the
ratchet to shorten the chain and tighten the load. It is a simple
enough concept, but it is new to me, it is physically difficult, and I
am struggling with it. So is Handlebar. When he gets one wrong, he
says, "Now they'll know a worm done it." When he gets one right,
he says, "Good enough for the girls I go with."

The word *worm* is an insult, of course, reserved for hands who don't know what they are doing. It can also be used, like the word *rigging*, as a verb, or it can refer to the thing that was done wrong, like a hook. So, if a greenhorn incorrectly attaches a hook you could conceivably say, for instance, "That worm wormed a worm."

Handlebar and I finish tying the derrick off to his trailer. He climbs back into his truck and rattles off.

FINALLY, THE RIG is completely disassembled, all the pieces on their way to the new end. After a full day of humping through trucks, forklifts, and cranes—with the sounds of pistons, hydraulics, metal screeching and grinding, and shouts of men over motors running—the emptiness of the landscape and the quiet of early evening appear sudden and brilliant. It is a shock to see this place of toil so immediately deserted—a peace settled in a seeming instant. Just off location, a few haul trucks sit quietly on the side of the road, still as a photograph. Dragonflies buzz past. Lizards in the dirt. For the first time while working, I watch the breeze ripple through the tall grass of the prairie that surrounds us. Mile upon mile upon mile of it stretches into the distance, teeming, clicking, and buzzing with thousands of species of life. The sun sits at the edge of the horizon, the day taking on a burnt-orange hue.

But the day is not over yet. The crane must be completely rigged down. Its boom must be broken into two pieces and strapped together in a way that will make the crane roughly the size of a tractor trailer, so the operator can drive it down the highway.

The crane operator stands in the mud with me for a moment. "I hope they send another swamper over to help you, but I don't know if they will," he says, his eyes narrowed in impatience, "goddammit."

"I'll do whatever you need," I say. "You just tell me. I've never done it before."

He gets back in the crane's cab. I stand where he tells me to stand, and he lowers the boom. As twilight tickles the earth's corners, I take a moment to feel proud of myself. Standing in the dirt in my mud-streaked coveralls in the final stretch of day, I breathe in the clean northern air. I did good today. I watch the crane's boom—120 feet long—as it comes down with something not unlike grace. It looks like a genuflecting dinosaur, bending its long neck to the earth in supplication. There is a sweet, religious air to the movement.

"I can't fuckin' do this with you." The crane operator stands outside the cab, his voice rising in exasperation. "You just don't know what the fuck you're doing. You don't know what the fuck to look for." I watch him quietly. "It isn't your fault," he yells in an accusatory tone, "but I can't do it with you."

I stand silently, and wait, chastened, until Handlebar shows up. "Well, now there's two of us that don't know what the fuck we are doing." He says, "How's this thing work?"

The boom of the crane is held together by steel pins. The pins can only be removed by sledgehammer once the boom is lowered to just the right spot where leverage can work in its favor. Bob Olhouser shows up, peppy as hell, and immediately sets to swinging the hammer. Handlebar is right behind him, taking a few shots at the pins. They are old dudes, but they sure can swing. Olhouser says, "Let Youngblood take a shot at it" and hands me the hammer. I have barely ever swung a sledgehammer in my life. My body aches. My arms are noodles. I raise the sledge and take a couple cracks at the pins. I'm hesitant, unsure of myself again. A missed swing can land on the boom and scar it, damaging it forever. I'll learn this later when a crane operator forgets to engage his brake and drops the boom on a mud tank, but I can tell something is wrong at that moment. Olhouser is ready to take the hammer away from me.

"No," says Handlebar. "Give him a damn shot. He's got it."

I take a deep breath and focus. I bang the pin through.

THE ASYLUM

Jesse's uncle is sitting on the couch with his shirt off. His form-less belly rolled out over his shorts. Skip's mom, Tammy, sits at his side, with her hands gently kneading his sweaty shoulders. She shakes with Tourette's as she rubs him. I try to walk through the room without engaging, but Ernie quickly apologizes for his snor-ing. His words come out fast, like a little kid hard-pressed to admit an infraction before Dad can get his belt off. I brush him off with a light laugh, but he continues.

"I'm really sorry," he says. "I was probably snoring so bad be-cause I spent the last six nights sleeping outside. I bought some medicine. It should help. You can just kick me if I ever snore again. Just give me a good kick. I don't care. If you want me to, I'll sleep in the bathroom. On the tiles. I'm sorry. I can't afford to get thrown out. I can't go back to sleeping in the park."

I pay Champ my rent in cash. He and Rickie are outside on the patio. I tell them Skip's mom is giving Ernie a massage. "What the fuck is going on in there?" Rickie asks. "When I came home, Jesse was on the couch holding hands with her."

"Jesus," I say.

The last time I'd seen Jesse and Tammy together, Jesse and Skip had been scrolling through ads for hookers on the internet and Tammy kept interrupting the conversation. Skip tried to shoo her away. "C'mon, Ma," he said, "I'm trying to . . . will you just give me a little privacy here? I'm just. I'm just looking for a little privacy. Just go to bed, will you?"

But she wouldn't leave. Skip tried to ignore her as she stood there silently watching.

"I'd fuck her," Skip told Jesse, clicking on an ad.

"She's not real," Jesse responded.

"I bet she is."

"You think she is?"

"Yeah."

"Three hundred dollars! That's a lot of money."

"I'd fuck her for three hundred dollars."

"No!"

"Yeah."

Pause.

"Yeah, I would, too."

Moving on to a new photo . . . "Is she real?"

"Yeah."

"I wouldn't fuck her."

Skip's mom then interrupted. "I'd buy you a prostitute if that is what you wanted," she told Skip.

"C'mon, Ma! That's not appropriate. You shouldn't. Don't say that. That's not something I want to hear. It's not appropriate."

"Really?" Jesse said, looking away from the computer and up at Tammy. "That's cool. I wish I had a mom that would do that for me. I wish I had any mom at all."

. . .

STANDING ON THE BACK PATIO, I relate the story to Champ and Rickie. "It's a fucking asylum," I say. The three of us pull on our beers in unison.

"We like you because you don't talk to us," Rickie tells me. "You don't care about my life and I don't care about yours."

That night, when Ernie again rumbles the living room walls with his wet, open maw, I grab my bedding, stomp across the street in my underwear, and crash again in the back of the Blazer.

SHOULDER TO THE WHEEL

A disembodied voice comes over the radio, "We always back into location, Mike, so, if there is an emergency, we can pull right out." It takes me a moment to realize that the voice is directed at me. "So, you want to turn around," Bobby Lee says.

I fumble with the CB and awkwardly key it. "Copy that," I say.

The day has only just started, and I've already been yelled at for driving the wrong direction from the rig down location. "Take a left," the trucker told me, "go a mile, take a right, go another mile, take a left, go about two miles then take a right and go three miles, and the location is on your left."

"Um, okay," I said, and immediately got lost. But I'm on location now, so I follow Bobby Lee's instructions. I turn the truck around, and back into a space in a row of pickup trucks lined up against a trailer. I'm not used to driving a truck as big as this Dodge dually, and as I back into the space, *bam!* I slam against the trailer.

My mouth contorts into a grimace, and I shut my eyes. "Fuck," I say. I hop out of the truck and walk around it. Thankfully, the trailer isn't damaged and no eyes are on it. No harm, no foul.

It is an hour into the workday, and my street clothes are slathered in dirt. "You don't need to put them greasers on just to drive over there," the guys on the other end told me, before adding. "Just c'mere for a minute first . . ." They convinced me to start work in my own clothes. Within ten minutes my jeans and shirt were stained by rust-red muck. After that, I changed into my FR jumpsuit and went to find the truck pusher.

On Sidewinder Canebrake 103, Bobby Lee is the rig up truck pusher. He stands in the midst of location wearing his uniform of a long-sleeved snap-button denim shirt with a pack of cigarettes in each breast pocket, and faded blue jeans tucked into cowboy boots. He's got the company radio in one hand, a USA Gold cigarette dangles from the other. I'd spent the past several days with Bob Olhouser, and the two Bobs could not be more different. They are the yin and yang of oil field truck pushers. Where Bob Olhouser is tightly wound, high strung, and high volume, Bobby Lee is loose limbed and nonchalant. Olhouser has a square head and probably cuts his own hair. Bobby Lee could have been a movie star. Olhouser is loud: a laborer at heart, he swings hammers, swamps cranes, and crawls up rigging. Bobby Lee has the easy metabolism of a boss, comporting himself like Hollywood's version of a small-town sheriff. Indeed, Lee is a onetime lawman, having, in his previous life, arrested some of the men he now oversees in the patch. Sometimes during long workdays, he puts his cowboy boots up on the dashboard of the van as if it were a desk, pulls his dirty cowboy hat over his eyes, and sleeps. I would come to think of these two men, over the months that followed, as two sides of the same cosmic petroleum-minted coin.

I walk to where Bobby stands in the middle of location. Around him is a swirl of activity. He looks like he could be bird-watching. "You're working with Porkchop today," Bobby Lee tells me. "Don't put your fingers anywhere you wouldn't put your dick."

Porkchop is a big Native American swamper with gang tattoos on his arms and neck. His teeth are filed to points. He wears a scraggly mustache under his shades and hard hat. He introduces himself softly and tells me that the swampers take care of each other. He says there are a few guys who will help me out, like him and Huck.

I tell Porkchop about the snoring at the flophouse. I'm exhausted, I admit. "I feel like I'm starting the day a step behind."

"You'll get used to it," he says. "When I was in prison my cell mate snored." He tells me his wife cheated on him when he moved from Idaho to Williston. "Child support payments are killing me," he says. He is twenty-four years old.

The day before, I'd seen Porkchop talking to a driller among a crowd of roughnecks. The driller was in a lift with another guy, and his voice was warm and Texas friendly. He was goofing on an extended talking jag.

"You always gotta be aware of your surroundings, buddy." The driller said, "I'm always aware. See the sun over there. I always know where the sun is. You see me up in that rigging? I been up in that rigging all morning, buddy. I don't have a watch on me. You got a watch? Take it out. I tell you what. Ask me what time it is."

Porkchop smiled carefully. A small group of guys had gathered to enjoy the driller's brag. "What time is it?" Porkchop asked him, playing along.

"See that sun, buddy? I'd say it is ten twenty-three."

Porkchop looked at the watch. "Holy shit!" He looked at the guys around him, back at the watch, and back at the driller.

"What time you got, buddy?" asked the driller.

"Ten twenty-three," said Porkchop.

"See what I mean? Always be aware of your surroundings."

At the time, I assumed all roughnecks were like this. I'd be proven wrong, but that was later. The roughnecks on Sidewinder Canebrake 103 were some of the best I'd ever work with. It seemed like every single one of them could operate a forklift, tell time by the sun, and swing a hammer upside down a hundred feet in the air. They appeared military in their bearing, upright, formal but dirty, and impressive. Several of them were veterans.

Ever since Civil War general M. H. Avery, commander of a Union Cavalry regiment, rode directly from Lee's surrender at Appomattox to Pithole City to organize an army train of teamsters, military men have been a constant presence in the oil field. There are many similarities between the jobs. Both are hard. They involve travel to inhospitable places, ridiculously long, physically demanding hours, and a discipline and grit beyond what most men and women are willing to endure. Former CIA director and US general David Petraeus, on a visit to North Dakota in 2013, would compare the Bakken to a war zone.

Soldiers and oil field hands are both indispensable to the functioning of modern society. Arguably, they are drawn from the bottom of society's totem pole, coming from the rougher edges of the American Dream, and born into their jobs most often because their dads did it. America fetishizes our armed forces. In my opinion, this makes it easier to discount soldiers as human beings. If we really loved them, they'd get better benefits. All that said, the roughnecks on Sidewinder Canebrake 103 seemed to me like Special Forces commandos, the gold standard of a certain type of manhood, and I was excited to be around them.

We are in the backyard of the rig surrounded by a wide array of machines that deal with mixing mud—mud pumps, mud tanks, a mud and shale shaker, and a mud and gas separator. Each of these components is easily the size and weight of a fire truck.

The driller and a roughneck from his crew, called Big Country, are standing on a platform working to fit two large pieces of steel piping together. The driller has his arms around one piece, straining to pull it up. Big Country is on top of the other part using his body weight to align it. Both men sweat and curse.

"You're going to help them," Porkchop says. He grabs a harness from the crane and shows me how to put it on. Like my jumpsuit, the harness is too big. It is heavy and the straps hang awkwardly off my body. I climb to the top of the platform and tie off to it, clipping a carabiner to a metal railing. We are about 20 feet in the air—high enough as to where you'll get hurt if you fall.

The harness greatly restricts my range of motion. I move forward to help Big Country and the driller but I struggle to get a hand on the pipe. "You gonna help us?" the driller asks.

"I'm gonna try," I say. We situate our bodies so each man has a hand on steel. We bend our knees and strain our backs, but the pipe doesn't budge. "This lil piece of pipe ain't gonna whip us, is it, Big Country?" the driller asks his buddy.

Big Country shakes his head no.

They holler down to Porkchop, who signals the crane. The crane swings its line toward us, and we wrap a sling around the top piece of pipe, but the angle is off; the crane can't maneuver properly, and the pipes still won't line up.

Porkchop hands us up a pry bar. Big Country wedges it under the metal and heaves. He's a giant of a man, but it doesn't budge. So, Porkchop hands up a sledgehammer. The driller swings at the pipe, then Big Country. I'm half his weight, but I take a few swipes myself. The hammer clanks against the pipe impotently, the steel refusing to give in. The sun beats down on the backs of our necks. We stop to catch our breath.

"Getting beat by this fuckin' piece of pipe," the driller mumbles. He and Big Country disconnect their harnesses from the platform to free up their movement. I disconnect mine and soon Porkchop and the crane operator are screaming bloody murder in my direction. I am too new and too dumb to take risks, too much of a little wormy fuck. It is humiliating. I obey and tie off again, standing there uselessly as the driller and Big Country work. They can't get it. They take another moment to rest. We swallow down some bottled water.

"Where I gotta go to find any meth here in North Dakota?" the driller asks me.

"Oh," I say. "I don't know."

"Fuck this," he says. "I'm jest happy I'm headin' back to Texas in a few."

He calls for a forklift. When it arrives, it positions its forks under our platform and lifts us all up just a bit. I feel the world tilt beneath me. The piping inches closer. We reconnect the crane's line to the pipe, but even working with the crane and the forklift in tandem, we still can't get the pieces connected.

"Goddammit," says the driller. "This fucking pipe is trying to whip us."

As I will learn, the oil field is a series of whippings, hammerings, and beatings. Rigs are hit, battered, thumped, and struck. Men get pummeled, knocked down, and clobbered. The job is a battle. If we don't *beat* it, it's gonna *whip* us.

Later in the day, I'll be struggling to remove a small steel pin from a piece of hydraulic tubing. Big Country will see me wrestling with it and say, "You ain't gonna let that lil pin beat you, are you?"

"Fuck no," I'll tell him. If he hadn't been there, I would have said fuck it and given up.

"How long you been doing this?" he asks me, back on the platform.

"A few days," I tell him.

He nods. "You'll have it whipped in a couple weeks," he says.

"Fuck yeah, I will," I say. I believe it, too. Because if I didn't, I would have given up right then, walked to the van and taken a nap until heading back to the yard and going home forever. As it turns out, I was completely wrong. A couple weeks? For somebody else maybe. For somebody with a knack for it. But for me? It was delusional. I had months of ass whippings in my future before I'd even feel I could go toe-to-toe with the job. I'm not sure, even years later, that I can say I ever got it beat.

The only thing I had going for me in those first days and weeks beyond stubborn pride was a fear of getting stranded. I'd gotten hired, but I hadn't been paid. Every step I put in front of the other was a step closer to that first paycheck. Without money in my pocket, I couldn't afford to leave even if I wanted to.

"Gimme a goddamn welder!" the driller shouts, exasperated. When the welder shows up, the two men talk it over, and the welder agrees to create a new fitting so the two pieces will join.

Because everything on the rig must be connected. Shakers connect to mud tanks, which connect to the drill string through mud pumps by a high-pressure hose. Electricity runs from the generators to every electric-run component. The drill line is threaded through the crown block and traveling block at the top of the mast. Like the thigh bone to the leg bone, nothing is extraneous, everything is necessary.

A decade after Drake's Well struck oil in Pennsylvania, the first gasoline-powered internal combustion engine was attached to a pushcart. Model T Fords were falling off the assembly line forty years later. In 1912, Britain converted its warships from coal to oil. It was these American and British oil-fueled machines that defeated Germany's coal-burning military operations in WWI. Petroleum was later decisive in the Allied defeat of Axis powers in WWII, Nazi horses proving to be no match for American trucks. In the biggest battle fought on the Western Front, superior supply lines allowed Allies to hold the line at Bastogne. The Battle of the Bulge decimated the German military.

A decade after the war, President Eisenhower signed the Federal-Aid Highway Act into law, initiating the single biggest public works project in American history. The bill transformed the geography of the nation unlike anything before or since, not only profoundly altering the physical topography of the country—a spider web of roads replacing the unruly backcountry trod by Lewis and Clark—but forever transforming the way humans experience it.

The Corps of Discovery traversed the new nation by navigating their way up the continent's many rivers; after 1956, Americans would negotiate the country on rivers of oil, experiencing its natural beauty through the windows of cars: cars powered by the petrochemical gasoline, over roads built by the petrochemical asphalt. Oil became the boat *and* the river.

THE CRANE SWINGS what looks to me like a weird piece of metal into the air. The iron roughneck is maroon in color and covered by a black sheen of oil and lubricants. It's over 10 feet tall and 5 feet across at its widest point, but it weighs between 4 and 6 tons, making it as heavy as ten grand pianos if they were stacked one on top of the other. Held together by thick bolts and welded joints, the iron roughneck's wheels, rivets, and arms are built to handle pipe as it goes in and out of the bore hole. The crane operator lands it safely onto the drilling floor, and the roughnecks work to position it.

I've been paired with a tall, lanky rigger called Smash. "They don't call him Smash for nothing," Porkchop told me. "Motherfucker'll kill ya."

Smash is a quiet guy with broad shoulders and a square jaw. The word *SMASH* has been scratched into his white hard hat by a blade. He wears a tan long-sleeve shirt and tan work pants. Twenty-three years old, from Dubois, Wyoming, he hates the work but he is married, and he needs the money. "No work at home," he says, the usual refrain. We stand in the sun next to the crane in silence, chewing, sucking, and spitting sunflower seeds into the dust.

A gruff crane operator has taken over. The Gruff Crane Operator is a goatee, a gravelly voice, and a pair of rainbow-colored Oakley's. He gets word over the radio that the roughnecks on the floor are having a hard time docking the gear. He leans out of the cab of the crane, lets loose a dark brown sluice of dip, and tells us to go help those dumb fuckers.

Smash paces quickly across location and up the stairs to the drilling floor. I follow. Up top, a group of company hands struggle with the iron roughneck. They are huddled around it, groaning against the weight of the machinery under the blasting sun, grunting a chorus of curses. Smash finds a place among the men and starts pushing. Bobby Lee is there, standing off to the side, a radio in his hand.

"What's Mike doing up here?" he asks no one in particular. I barely register that he is talking about me. "What you doing up here, Mike?"

I get a hold on the metal. "I'm learning stuff, Bobby," I shout over the groans and curses. "I'm getting my hands on things, so I can get learned up."

We heave against the metal. I smell sweat and diesel, and I push. It doesn't budge, so we take a break. Smash moves to a position where the crane operator can see him, and he signals the crane with his hand. Thumb up like a hitchhiker, fingers folding in and out. Boom up and hold your load. We get our hands back on it, the roughnecks and I, and we dock the equipment. We unhook the crane and move on to the next task.

The day comes to a stop when some metal gets torn up, and welders are called in to make a fix. I sit in the van with the group of guys who share my job. It is Porkchop, Huck, and Smash, a muscular swamper with braces, and me.

Huck is telling the guys about his first strip club experience. "The stripper shoved her hands right down my pants and grabbed my dick!" he exclaims.

"Did she give you a blow job?" Porkchop asks.

"I could really use a blow job," says Smash.

Porkchop suggests that we give each other blow jobs to pass the time. He and Smash sit in the two front seats, and Porkchop says he thinks I should get between them and give Smash and him simultaneous hand jobs.

"It's all in the elbows," he says, and demonstrates.

I am the oldest guy there by over a decade. I feel out of place. These guys are kids, I realize. At the same time, they know their jobs better than I do, and I need their help if I am going to make it. Fuck it, we rock the van back and forth by throwing our collective weight from one side to the other in the hope it will look to people outside like we are all fucking each other.

The air conditioner in the van only works when the vehicle is moving, so Porkchop puts it in gear, and we cruise in a circle around the rig. But the van starts kicking up dust and annoying the guys who are still trying to work, so Bobby Lee comes over the radio and tells us to drive it off location. Porkchop steers the van out onto the access road that leads off the job site, and we rumble into the surrounding prairie.

We enter into a vast landscape of gently sloping hills rolling into the distance, unobstructed by mountains or trees. The grassland tilts to the breeze, rippling in gentle waves washing easy over the undulating curves of land. The prairie-ocean. Here and there a cell tower or another rig or pump jack can be seen on the horizon, but with meager exception the visible earth is free of man and commerce.

We cruise down the oil field roads and farm lanes of the prairie kicking up a comet tail of dust and pull onto a location wellhead where two pump jacks swing up and down in the herky-jerky motion of rocking horses. We leap from the vehicle, and Huck picks up several large chunks of scoria. He throws them at the pump jack and they smash into puffs of red dust. We cheer. Smash and Huck tell us they like getting drunk, driving around at night and shooting the pump jacks with handguns. We climb back into the van and cruise back to location.

The usual length of a workday on a rig move is between twelve and fourteen hours. On this day, my first week in the patch, we will put in seventeen and a half hours of work.

AT THE END OF IT, we leave location and Bobby Lee drives us to the closest truck stop, a big, brand-new corporate establishment just south of Tioga. He says he will buy us burgers, pizza, soda pop, whatever we want. He says it's on him. There is a trace of paternal pride in his soft voice. We climb out of the van and push through the doors of the truck stop coated in dirt and dust, invert, diesel, and mud. It feels good swaggering into the truck stop with this group of rough-and-ready men, a truck pusher and his crew of hands, guys who put their shoulder to the wheel of the world, push it, and make it turn.

A SURPRISE

We land back at the yard at 11:30 p.m. Bobby Lee tells us we need to be into work the next day at 6:00 a.m. For a moment, I think he is joking, but I keep my mouth shut. The other guys are quiet and thoughtful as we collect our lunch box coolers and climb out of the van. I've been up since 5:00 a.m. It is only a fifteen-minute drive back to the flop, but I am wired. My brain is turning fast; my whole body feels like something inside it is spinning. The week has been a testosterone-fueled adrenaline rush. By the time I get home, it will be midnight. It will take time to slow down, and I'll have to be up again at 5:00 a.m. to do it again on less than four hours' sleep.

The yard is dark. We make our way toward our respective vehicles. I pull myself into the Chevy and feel my body weighing heavy on my bones. The other hands pile into their cars and drive off. I take a breath and rub my eyes before turning the key.

It is the second of July. Folks all across America's boomtown are shooting off firecrackers. They have been all week. I pick up a six-pack at the Kum & Go, and when I reach the flop, instead of walking through the front door, I unlatch the fence to the right of the house's entrance and walk around to the back patio. I take a seat at the round metal table and open a can of beer. I find comfort in the familiar pop of the tab and fizz of carbonation. I take a long, cool swallow. Roman candles whine and explode, blackjacks pop, and moon rockets whistle in the distance. My body feels drained by work, my skin dry and burned from wind and sun. I soak in the sound and the occasional flash of firecrackers as they splash against the purple curtain of night. I sit inside a good empty feeling.

The back door to the house opens and a large man with a long gray ponytail and a drawn and weathered face steps out of the door. He is shorter than I am but with shoulders nearly as wide as he is tall. His eyes scan the backyard until they settle on me. He approaches.

He tells me his name, and I shake a big, calloused dry hand. I give my name. He nods, sits at the table, and begins hand rolling a cigarette. I sip my beer for a moment. Then I get up without a word and walk into the house.

I don't have time for this shit. I am too fucking tired, annoyed that Jesse and his uncle would invite a friend over so late at night. I want to give them hell about it. I step through the kitchen, past the sink full of dishes and the table covered in mail, empty beer cans, and dirty clothes. I walk past the laundry room with the brown-stained toilet and into the living room. They are all in their usual spots. Uncle Ernie sprawled out on the couch, Cholly lying in the corner on his mattress, and Jesse in his dirty oversized white T-shirt on an air mattress in the middle of the room. Their faces, like the faces of sad robots, lit by the flickering image of the large-screen TV, are blank. My small mattress, crammed in a corner on top of a dirty rug on top of the stained carpet, is empty. The coffee table has disappeared, and a new air mattress has taken its place. I realize in a blink of an eye that the guy outside is not anyone's friend. Champ has taken on a new boarder. The small three-bedroom townhouse I live in now holds twelve adults. There will be five of us sleeping in the living room.

I drop heavily onto my mattress and stare blankly at the TV. I finish my beer with a swallow and set the empty can on the dirty carpet next to my bed. I fall asleep in my clothes.

THE THIRD OF JULY

Champ is talking in torrents, gesticulating with a large metal spatula. Behind him, burgers sizzle and pop on the backyard grill. His shirt is off, and his shorts are slung low on his hips, his large white, red-haired belly sways in front of him as he mouths on, stepping from foot to foot like a doped-up boxer. He stumbles once, nearly falling back onto the grill before regaining his balance. He doesn't stop talking, though.

After work, I walked in through the front door of the flophouse to see two sets of bunk beds set up on either side of the living room. Jesse and Ernie sat on the couch watching TV. "Champ's outside," they said. "He wants to talk to you."

I didn't want to talk to him. I stepped into the bathroom and took a look at myself. I was covered in dirt. I took off my John Deere trucker's cap. My hair was matted with sweat and dust. My face and neck were burnt mahogany, but my forehead and the skin around my eyes were alabaster white.

I'd been texting with Champ all day. I told him I wanted a break in the rent. When I'd paid up two days earlier, there had been four guys sleeping in the living room. Now there were five. In my mind, that changed the math.

I had a good relationship with Champ. I was sure we could figure something out. Really, all I wanted was a hundred bucks. But it wasn't even just about the money. I needed my dignity. I had felt so many things slip away from me over the past month that this took on a greater, somewhat symbolic importance to me. Champ asked me if it would change anything if he bought bunk beds, and I grew tired of texting. I decided to ignore him for the rest of the workday. Now, I was home. I took off my boots and grabbed a beer from the fridge.

"You're my hero," said Rickie. She had come in from the backyard and shut the door behind her. I was standing at the fridge. Rickie was sweaty, her bangs stuck to her forehead, she talked in a rush. "I told Champ, I keep telling him, it's getting to be too much. There are too many people in the house," she said. "Things could get violent. But he doesn't listen to me. I don't pay rent. Someone needs to stand up to him."

I was taken aback. I didn't want to stand up to him. I didn't think I needed to. I didn't think it was a big deal. But Champ did.

He waves the spatula at me as sentences pour out of him in disjointed repetitive waves: "You have a legitimate gripe," he tells me. "I know I fucked up, but I got bunk beds to address the issue. I'll buy you a twelve-pack, dude. I need to make my money while I can. Those bunk beds cost me eight hundred dollars. We like you, dude." He tells me that he pointed me in the right direction to get my job. "But let's face it, you aren't going to stay around here now that you're making real money.

"I'm helping people. I'm giving people a roof over their heads, bro. Otherwise, they'd be fucking homeless. I know I'm a slumlord," he says. He tells me he won't give me a break, but that if I leave right now, he'll refund the rent in full. "At least I'm helping people," he says.

"Well, you got me by the balls, Champ," I tell him. "If you aren't going to budge, then I don't have any options here." I feel the tension between my shoulder blades, the twitch in my left eye, and the return of that familiar gut squeeze of poverty. I have no leverage beyond a handshake agreement that I thought existed between us, and he is breaking that.

Champ flips the sizzling meat and suckles his beer. "I'm a fucking slumlord, dude. Okay?" He shrugs aggressively, "I'll say it." His face is beet red, his forehead in a twist. He stumbles as he speaks, and he keeps talking, gripping tight to the metal spatula, his eyes small and cruel. He tells me that he'll give me forty-eight hours to decide if I want to move out. "A grace period," he calls it. He motions to Rickie, to include her in the conversation.

"I think he has a pretty good point, Champ," she says.

The high timbre of her soft voice hangs in the air for the longest of brief moments. Champ's body freezes in time, while a look of disbelief, then horror, then righteous belly-deep anger crosses his face fast as pictures in a flip book. He looks at her. He closes his eyes. He looks back at me. "Will you leave us for a minute?"

I walk back inside.

Jesse and his uncle remain on the couch staring blankly at the television. I beckon Jesse to sit with me on the front stoop. "I'm really stressed out, man," Uncle Ernie says as we walk outside.

"Champ spent eight hundred dollars on bunk beds," I say to Jesse. "He could have just not gotten another boarder."

As if on cue, the new boarder stumbles out the front door with a can of Natural Light in his hand. He has an alcoholic's face, porous and scarred by what looks like a combination of childhood acne and knife fights. He stands silently in the front yard, staring at the house across the street in a treacherous squint. A family can be seen through the front windows. Hopefully, he doesn't murder them, I think.

The night previous, Jesse witnessed the new boarder's night terrors. He was punching and kicking and grunting in his sleep. "I don't like that," says Jesse.

"He smells like prison," I say.

The front door opens, and Champ leans out. Veins pop from his neck and forehead. He squares his feet and squats down, his body coiled like a snake. He extends a thick arm out toward me, a cluster of cash in his meaty hand.

"Here it is, dude," he hisses. "If you wanna fucking move out, here's your fucking money. You want it, you take it. Four hundred fifty dollars, dude."

I don't budge. "I don't want the money, Champ. I told you I'm staying tonight."

"You have forty-six hours now, okay?" He moves back inside, slamming the door shut behind him.

"Jesus fucking Christ," I say, "another Williston curveball."

Ernie pokes his head out the door. "I'm really stressed out, man," he says. "I can't find the beans for the burritos I'm making." His voice is quavering. He looks on the verge of tears.

Back inside, I claim a bottom bunk and lie down to read, to pass the time. Life in the flop is how I imagine jail must be on work release. It's all about killing time, watching TV, maybe reading, trying not to let the other guys get on your nerves too bad. So you can get back to sleep. So you can wake up. So you can go back to work.

Jesse and Ernie eat burritos on the couch. The new boarder sits

on the floor, drinking. Champ bursts into the room all but spinning like a cartoon Tasmanian devil, his red face turning purple.
"This is my fucking house," Champ roars, pointing at his chest. "I'm
a fucking slumlord, okay? This isn't a fucking democracy! This is a
fucking dictatorship! Either you do it my way, or you get the fuck
out! Is that understood?" He storms out of the house and into his
truck. The engine turns, and he takes off down the road.

"In all of my shit-fucking life," the new boarder proclaims, "I've
never lived in a place like this."

Huh, I think. *I have.*

Champ returns not long after. He walks into the living room and
drops an eighteen-pack of Budweiser on the floor. "Here you go,
dude," he says to me. It is his peace offering. I stare at it for a moment before getting up, opening the box, and grabbing a beer. Fuck
it. I take a long, cool swallow.

That night, both Ernie and the new boarder snore loud enough
to shake the walls. I gather my pillows and a blanket, and I walk out
across the lawn and into the back of my SUV, where I try to get
some sleep.

INDEPENDENCE DAY

The Jamaicans weren't regular drinkers, but one night, when no one else was around, we shared a few beers in the living room. Daniel nodded along as I played guitar, and Gabriel and Cholly told me about Jamaica. They told me where I could find good hiking. They mentioned it a couple times.

"You guys do a lot of hiking?" I asked, and they looked at each other sheepishly.

"No," Gabriel said finally, and we burst out laughing as soon as his meaning became clear: white people always ask about hiking.

Cholly requested a Bob Marley song. "No way," I said. But they began to goad me on, and I acquiesced. I played through the first chords, feeling somewhat silly, but slowly working the changes. They listened attentively as, unsteadily, I warbled the opening lines. "I remember when we used to sit . . ." Cholly laughed and Daniel clapped his hands and all three joined in, helping me through the first verse and culminating in the four of us belting out the chorus, "No Woman, No Cry"! Cholly told me that Georgie, who, in the song, *keeps the fire lit*, was still alive. "He has huge dreads," Gabriel said. "And he still hangs out in Trench Town." We sang the whole song. We sounded terrible, and it was fantastic.

I had been looking forward to the Fourth of July. It was to be my first day off work since I started in the patch, only fourteen of the longest-days-of-my-life ago, and it had grown to outsized importance in my mind. Gabriel and Cholly were to have the day off, too. It was so rare, because of our upside-down schedules—they worked nights in hotels and gas stations—that we got to spend any time together, and I wanted it to be special.

For the Fourth, I imagined picking up chicken and sausages, vegetables and beer, and firing up the grill. Most of the people in the house would be working, so we would have the backyard to ourselves. The Jamaicans reminded me of New York, the only place that had ever felt to me like home, and I thought that I could bring that sense of home to Williston. Maybe this would make us real friends.

It isn't to be. I need a new place to live. I wake up and immediately look online for a room to rent. I'm down to just over $400 in cash and $600 in credit. Not good. But I have a friend back east who I think I can borrow some money from. If Jesse and I shared a $1,000 room, we'd be paying only $50 more monthly than we are at Champ's. Instead of five guys crammed into one space, there would just be two of us. In my pre-Williston life, I never would have considered sharing a bedroom with a nineteen-year-old kid I'd known for a month to be *a good deal*, but things have changed.

Jesse says he is on board. We drive to Lonnie's Roadhouse Café. Lonnie's is a diner slapped on an uneven dirt parking lot just off Route 2, connected to the entrance of Walmart. Not long after we eat there, a woman will steal a handful of bras from Walmart, accidently shoot herself in the parking lot, and walk into Lonnie's, where she will bleed all over the bathroom until the police come and pick her up.

The restaurant is built around a U-shaped counter. We take our seats in a booth in the far corner. Along the wall are old posters, sayings, Cracker Barrel–type bric-a-brac. We sit under a tin sign advertising "Picaninny Freeze," a watermelon-flavored candy. The ad is a racist drawing of a little black girl with bright red lips, buck teeth, and wide white eyes sucking on a watermelon. I've never seen anything like it, not in public anyway—and I experience a sharp moment of dread. I don't want to sit next to this fucking thing! I look around the room. It is all white working class, of course. I grunt something about the sign to Jesse, and he just shrugs. I try to forget about it, but the image stays there throughout breakfast, peering over my shoulder.

We order our food, and I tell Jesse that if we find a room share, I can help him get a job at Diamondback, and he can pay me back when he gets paid. He sips his coffee and raises an eyebrow. "Sounds good to me," he says.

It feels good to take charge. "I'm certain we can make it work," I tell him.

We drive back to the flop. Jesse heads inside, and I call my friend. I ask him if I can borrow some money, just for a few weeks. He hesitates but says okay. It will only be for a short time, I assure him, I'm in a pinch. I don't know what else to do.

Hanging up the phone, I realize that I am incredibly lucky. The ability to ask someone for a small loan gives me a tremendous leg up over most of the people I have met in Williston.

. . .

SEVERAL HOURS LATER I find myself sitting in a parking lot in a state of complete despair. "I think my bad luck is starting to rub off on you," Jesse tells me. I look at him but say nothing. I think he is right.

The first place we'd looked at that day was spacious and clean with a large open kitchen. It smelled wonderful, like vanilla and babies. The renters were a soft-voiced young hairdresser and his beautiful wife. After saying goodbye to them, Jesse and I trotted down the stairs to the parking lot.

"I feel like we just walked through a portal to some strange dimension," I said.

"That wasn't like Williston," Jesse replied.

"It wasn't like Champ's," I said.

"Fuck no." We hopped inside the Blazer. "He seems real nice," said Jesse.

As we drove to our next appointment, I told Jessie I was surprised at just how much Champ had freaked out over the rent.

"I'm not surprised," Jesse told me. Besides working at Pepsi, he had been picking up shifts at a grocery store. "I was a couple days late with rent and Champ showed up at the store and started yelling at me. 'I'm gonna get my fuckin' money! You better give me my money!' He was acting crazy, man. Nearly got me fired. When I got things settled, he calmed down. But now I gotta float Ernie. It's stretching me thin, man."

The second room we saw was small and cramped with ratty carpeting and a single window looking out onto a brick wall. The renter was a lonesome man who followed us around the apartment cradling his giant gray cat, a Maine coon, like a child in his arms. He spooked me.

As we left his place, I asked Jesse, "What are we waiting for?" I called the hairdresser. He said we could move in that day. So, I drove us back to Champ's and when Jesse went inside, I arranged the loan from my friend. He offered me $1,200, enough to cover the first month's rent plus $200 to buy two mattresses for the floor.

Inside the flop, Uncle Ernie was dejected. He sat on the couch helpless as a baby. "Champ is gonna be pissed," he said. "He's gonna kick me out."

Walking out of the kitchen, Gabriel flashed me his wide, slow smile and I returned it. "Mikey," he said. We clasped hands. I told him I was moving out and felt a sharp pang of regret watching his smile disappear. I really wanted to spend the day with him. It seemed bigger than us somehow, my desire to grill and sing with the Jamaicans. I had thought that on this day, with them, I could create a vision of the America I aspire to, a place where even in a flophouse on the edge of the world, a motley crew of immigrants and migrants could choose sharing, choose strength in numbers, choose cheeseburgers, music, and friendship. But it was too late. I had to go.

Jesse and I loaded our stuff into my SUV. Everything we had in the world fit in the back of my Chevy Blazer. Just as we started down the road, I got a text message from the hairdresser. He apologized. As quick as we had a room, it was taken away. He gave no reason. I could only guess that his wife vetoed his decision.

We now sat in a parking lot, stunned. "Another goddamn Williston curveball," I say. At this point, it feels like we're striking out.

"Fuck it," I tell Jesse. I'm not going to let this beat me. This little thing ain't gonna whip me. I get on the phone. I call every single place I can find online that is renting a room, I leave message after message, and we drive to the first place that responds, on the outskirts of town. A small house on a hill. A man leads us inside. Dirty. Run-down. No TV. No internet. $1,500 a month.

It is late afternoon now, the Fourth slipping away. "I need a place to think," I tell Jesse. I pay $100 for a room at a motel next to the Walmart. The parking lot is full of beaters with out-of-state plates. We can't leave our stuff in my vehicle, we'll get robbed. It takes several trips to carry our belongings into the motel room.

I am walking down the hallway with an armload of stuff when I hear, "Yo! What up, B?!" I turn to see Frank poking his bald head around the corner. He swaggers toward me in his Adidas track suit. "What up?! Yo, this is my B." He introduces me to his friend, an imposing light-skinned dude wearing a tight T-shirt.

Frank wants to know why Jesse and I are there. Jesse fills him in.

"Champ is a meth head. I had to get out of that fucking place," Frank tells us. We ask what he is doing at the motel. "We're hanging upstairs, B! Talking to bitches on the internet! I got all kinds of bitches on the internet. Thai, Filipino. Bitches are sending me naked photos and shit!"

When Jesse and I get back to our room, I turn to him. "Whatever you do, do not give that fucking guy our room number."

We lie down on our motel beds, and I call a woman I had been dating in New York. It is good to speak with someone kind, funny, and caring, but I end it quickly. I can feel the conversation opening something up in me, a sensitivity that is of no use to me now.

I swallow my way through a beer and listen absently as Jesse talks to his daughter. Lying on the bed with his knobby legs pulled toward his chest, he holds a flip phone to his ear. His left arm, skinny as a snake, is outstretched, toying idly with an empty can of chewing tobacco. He takes up the whole bed and none of it at the same time, a stick figure in baggy jeans and T-shirt. He speaks to his daughter in a high-pitched child's voice, "You going to fireworks? You gonna hear the bang today?" he asks, a kid talking to a baby. I feel embarrassed listening, invasive. But there's nowhere for me to go.

I keep turning our predicament over in my mind, trying to see it from new angles, trying to develop some sort of plan. I feel responsible for Jesse. This was my idea. I dragged him out of the flop. Even if it was a shit show, it was a roof over his head. I turn the TV on, then I turn it off.

When Jesse finishes talking to his baby and her mama, he calls Uncle Ernie. Ernie tells him that Champ has apologized to everyone for his behavior. I take a breath. I call Champ. Champ asks where we found a place, and I tell him the situation. He says if we come back, he will cut us a deal. I feel the sharp tension in my back, the twitch in my eye, the reckoning with the Williston curveball. Jesse and I discuss it, but it doesn't take long before we reach a conclusion. We have no other options.

RICKIE SITS AT the patio table with puffy eyes.

"How are you doing?" I ask her.

"I'm here," she says.

Champ is jittery, his eyes unsettled, his face hanging off his skull as he works the grill. He is half-apologetic, but still talking circles, still combative. He says he knows he was an asshole, and I take it as a chance to lighten the situation. "You really were an asshole, Champ," I say. Jesse and Rickie start to laugh but Champ does not, and they both stop themselves.

"You don't need to say that," says Champ.

He gives me my rent money and tells me I can stay for $15 a

night. As soon as I find a new place, I am welcome to move out. I thank him. He heads inside, and I look at Jesse. He takes a deep breath, puffs out his cheeks, and exhales.

Finally, the sun is setting. Several of the others slowly wander out into the backyard. Skip, Tammy, Uncle Ernie, and the new boarder. We don't talk much. Just listen to the fizzle and pop of firecrackers sizzling their way over America's boomtown, slapping against the weight of the sky. Two drunk men in flannel shirts and ball caps stagger down the alleyway and set up Roman candles. They light the fuses with cigarettes and hastily back away. The rockets sputter, then take to flight, climbing upward in an elegant arc. They splash color against the night. It is Independence Day.

"What a waste of money," says Skip.

RIDE

Huck wings an empty bottle of Bud Light Platinum out the window of the work van. The bright blue glass spins in the light of the lowering sun and disappears into the switchgrass blurring past the windows. Porkchop has Bobby Lee's van at full tilt, burning ninety miles an hour west on Route 2. We are headed home at the end of a long day. Huck twists the top off another bottle. He brings it to his lips and sucks half of it down in one long, cool draw before tucking it between his thighs.

"Ahhh, yeah. Ole Pops asked me if I'd ever thought about killing someone. 'Uh, yeah,' I said, 'I've thought about it. I guess.'" Huck lets loose his goofy laugh. "He's got all kinds of guns, crazy old fuck. He told me he had some unmarked, untraceable guns, and if I ever needed one, he'd give it to me, no questions asked. I said, 'Uh, thanks.'" Huck shrugs. "I mean, I guess," he adds doubtfully.

WE HAD JUST finished a rig move. I was swamping for a gin truck driver called Pops, the meanest, most miserable bastard I'd ever met in my life. My first morning with Pops, he spoke to me with grandfatherly assurance. He said, "It's my policy to not make you do anything I wouldn't do." He wagged his finger at me and smiled. "I want all my swampers to finish the day with all their fingers and toes. Safety is the most important thing."

"HOW WAS IT working with him?" Huck asks me.

"He told me to get out of the truck while it was still moving," I say, "and dropped me in the middle of a big, swampy puddle when there was dry ground just a few yards away."

Huck and Porkchop exchange a glance and a smile. "Did he get to cussing at you?" asks Huck.

"Fuck yeah, he got to cussing at me. He turned purple if I got something wrong, and if I got it right then I was moving too slow for him. 'No! No! Goddammit, no! Do it this week!' He cussed me no matter what I did," I say.

Huck and Porkchop laugh out loud, and it lightens my mood. The four days with Pops were awful. I felt like I was carrying fifty-pound weights on my chest, my stomach in knots, my brain yelling at itself. The feeling hadn't completely gone away.

The night before the rig move it rained, and I made the mistake of thinking I might get the day off. Like a grade-school kid on a snowy night, I couldn't shake the thought. It kept me up, tossing and turning all night long. By morning, I was more tired than when I'd gone to bed.

I knew I was in trouble as soon as I climbed into the truck with Pops. Like a chicken in a pen, the old man could sense a speck of blood on me. It disgusted him.

I was a step behind all morning. I couldn't seem to get my hands on anything. I wasn't asserting myself in situations where I understood what to do, and I wasn't asking questions quickly when I didn't know what to do. I was showing up late, standing by and watching. I was the worst thing a field hand could be: I was in the way, and I knew it.

"Hey there," a Texan bed truck driver yelled at me from his rig. "Are you wanted by the FBI?"

"No, dude, I'm not," I yelled back at him.

"'Cause I ain't seen you get your fingerprints on a goddamn thing," he said.

I thought about quitting. It wormed its way deep behind my ear and squirmed into my brain until all I could think about was giving up. I imagined planting a white flag in the middle of location and lying down in the golden peace of sweet surrender.

But I didn't. I put my focus exclusively on each immediate task. I completed one thing, then I moved to the next thing. Moment by moment, I inched my way through an endless, angry day.

"Hurry up, you dumb cunt!" Pops screamed at me.

It will only get easier, I told myself. This is as hard as it can get.

"I'm doing the best job I can," I said, confronting Pops after a morning of insults. "You're yelling at me whether I'm doing something right or wrong. It isn't helping. You've got me so I'm fucking up things I know how to do."

After that, Pops wouldn't talk to me or look at me. I told him I was going to help some guys who were moving mats with the crane. "Beep the horn if you need me," I said.

I stood on the mats and when the chains swung to me, I took the hooks in my hands and slid them into the channels in the mats' four corners, and I watched them fly out. Simple and satisfying, it allowed me a moment to breathe, to get my bearings, to calm down and regain myself.

Across location, I saw Pops climbing out of his truck to rig a piece of metal by himself. He'd been sitting on his ass his whole life. He was old and fat, stooped and bent. I felt a surge of sympathy and made my way over toward him, but he waved me off dismissively. He didn't want my help. I was angry, disappointed in myself, humiliated, but also relieved. *Fuck him*, I thought. I returned to the mats and spent a few glorious hours being useful.

At the end of the day, when I climbed into the truck, Pops was back in grandfather mode. He told me about a swamper who jumped between his truck and a rig up. He was attempting to adjust the chains while Pops was hoisting an object into the air. This swamper was hit so hard he dislocated his shoulder. "The ride back to location was over these bumpy dirt roads," Pops grinned. "By the time we got back to Diamondback, his shoulder had already popped back in." He let out a low, friendly chuckle and turned to me with a twinkle in his eye.

On Sunday, a brief respite came when Pops turned the radio to a local church broadcast. We sat in the truck and listened to a Protestant choir sing "He Is Risen" while watching the rig come down in the mud.

IN THE VAN, on the way home from the completed job, Huck and Porkchop get a kick out of my story. It lightens the mood and puts me at ease. They both had to endure the same day I just endured. Many times. It is part of learning the job.

"You just can't let it get under your skin too bad," Huck says. "It'll get easier."

He finishes another beer and sends it sailing out the window, electric blue helicoptering against the dusk, slipping into the smear of prairie. He passes a full bottle back to me, and I take it, letting a swallow rinse my throat before tucking it between my thighs.

Huck had pulled the wad of dip from his mouth and flung it out the window when he started drinking. Now, he lights up a cigarette. "I smoke until my lungs hurt, then I dip until my cheek hurts, then I smoke until my lungs hurt," he says happily. "I do that all day long. Helps pass the time."

It seemed like it did. For all the toil of working the patch, there was a lot of standing around, too. Tobacco was the only form of debauchery on hand for a man to partake in. Nothing kills time like a vice.

"You mind if I get a pinch of that?" I ask.

Huck shrugs. "Sure," he says, passing back a can of chew.

I open it up and stuff a dollop of the wet, threadlike tobacco under my lower lip. It stings like a strong spice. It feels awkward against my gums, the taste sickly and sweet. I hold on to it, but I stop listening to Huck and Porkchop. They are talking just to pass the time. I feel the juice sliding back into my mouth, and I have to remember to spit it into an empty bottle. A minute later, a few more spits, and I take the dip out of my lip and throw it in the trash. Huck and Porkchop laugh.

"Well, now you know," Huck says.

PORKCHOP TELLS US he has had night terrors ever since his last stint in prison. In prison, he heard guys getting their throats slit while they were sleeping. Once he saw a guy jabbed in his jugular in the cafeteria and killed. "Sometimes I wake up feeling like I can't breathe," he says. "I wake up and check around my neck for blood."

"What sent you to prison?" I ask.

"I killed a guy." He smiles and I see the filed points of his teeth. He gestures to the tattoo on his right bicep. Two hands together in prayer with a nail driven through each hand. "It's a Blood tattoo," he tells me. When I press him, he says the nails represent the two things he did to get initiated into the gang. "They led to my time," he says. As he drives, he lifts up his shirt, revealing a deep, nasty scar across his belly. "Knife wound," he says.

I'll learn later not to put much stock in Porkchop's stories. But that evening, sitting in the van with the prairie blurring past, I believe him. He is a man who needs the approval of other men. He hungers for it. I see that in him. And I can imagine him killing someone because he is told to do it.

I let my bones sink into my seat. *If Porkchop is a killer,* I think, *what does that make me? A friend of killers?* I like Porkchop. I think he's great. He helps me. I am coming to rely on him. Some guys kill other guys. It isn't right, but maybe it isn't my business, either. Is this completely nuts? Should I confront him somehow? I've never had to think about this before.

In Williston, I would be confronted by this again and again. Not only on this lazy evening drive but later, in starker terms, when I meet Erwin "Jack" Jackson—a man I would become truly close with, and a man who had committed heinous criminal acts. I struggled with this question even as I enjoyed the company of unabashed bigots and learned to compartmentalize their casual, constant, continuing faucet drip of racism. How terrible is that? Does that make me a bad person? I don't know. How do you love men you disagree with so violently on the ethical and moral questions that you think define you?

Later, the questions I pose will kind of invert themselves, as I turn them toward toward the world. Because how do you *not* allow yourself to love people you disagree with? Wouldn't that be a sign of real cowardice?

The sun still sits on the horizon as we pull into the yard. It is simply refusing to set, the sun, refusing to pull the day down into a calming, cool night. Behind us, over a stretch of sixty miles, twelve empty blue bottles line the roadside.

HUCK ROLLS SOME TRUCKS

A regular topic of conversation among swampers was car wrecks and DUIs. They were sources of pride for a lot of guys. Huck got his first DUI a few days before Christmas the winter before I met him. He'd "piss pounded five tall Amber Bocks" in less than two hours at a Chinese restaurant in Williston and then driven out to his parents' trailer. He wasn't yet twenty-one, but before getting his DUI, he'd already rolled more than one truck.

He wasn't the only swamper who lived like this, but Huck could tell the stories better than any of them. He was an instinctive, animated raconteur. I almost got the sense that he got in trouble just so he could talk about it later. Storytelling brought out his innate joy and contrasting melancholia, his goofy sense of self but also a detached, critical eye that I only saw otherwise when he talked about the internal workings of motors. He had a wry sense of humor, an explosive laugh, and an uncanny knack for finding that odd detail that made his stories crackle. I would witness a lot of guys tell Huck he was stupid. They were wrong. Huck knowingly portrayed a role when he spun his tales, and he used that role—the dumb, young, full-of-cum oil field hand, out to raise hell and trash the town—to make his narratives not only funnier, but sharper, more perceptive, more a part of the fabric of the boom. He was a natural.

"You had five Amber Bocks and you're driving home," I prompt him.

"I get outside of town, and when I come off of Highway 2 turning onto County 8, I fucking just gunned it. I was doing about seventy by the time I got to the top of the hill, and the cop was right there."

Huck makes the sound of a police siren and it sounds just like an actual police siren. "That's pretty good!" I say.

He nods and continues, "Seventy in a forty-five. I governed out my truck when I hit dirt. There's this crossroad, County 9, and it's kinda lifted a little bit—I hit that shit at like a hundred miles an hour and I caught air. I don't know how I kept hold of it. Pickups aren't aerodynamic! The wind is going under it." He smacks his right palm against his upturned left palm showing the truck planing across the crossroads and making a sound like *Fssss Fsss*. "Doing a hundred, going side to side! I was able to get outta that. But the cop still got me."

In Huck's estimation, receiving a DUI started him on his year of trouble, trouble with driving, trouble with drinking, trouble with women, and trouble with the law. "I'd been working like a dog," he would tell me. "I worked for twenty-six days in a row before I had like one or two days off. And then back on for like a month. And it was like . . ." He puts a finger to his head and mimes a gunshot. "Kill me, please. And I was partying like crazy. I just kinda lost my give-a-fucker."

He had already rolled his 2000 Dodge Ram three-quarter-ton pickup. "It had a diesel in it and a six-speed manual. It was a machine, dude. And I was hauling ass. Drifting around. I hit this turn, and I didn't really slide. So, I backed up and went and did it again. I got her sideways and I was like 'Fuck yeah!' and then I just gunned it. Got it up to like eighty. The tire completely came off the wheel. It was just like *POP!* And then the Ram hit dirt. And that's when I started rolling."

Huck is up on his feet telling the story, his whole body bent into it. He feels he misses a detail, and it is like watching a human VCR quickly rewind and then fast forward through a scene. His voice becomes Alvin-and-the-Chipmunks-fast as he gushes, "But how it started is I was just hauling ass and I started kicking the wheel. I started kicking the wheel to one side and it's like . . ." Here Huck makes the sound of gravel kicking up. "Then I'm like, 'Oh fuck!' I started sliding sideways at a high rate of speed and that's when the wheel came off and SNAP!" Huck's girlfriend at the time, Jenny, was riding shotgun, begging him to slow down. "I think we rolled like four or five times. Jenny got shot out the window. She went into the ditch, and I just kept rolling. She fractured a clavicle and tore some muscles in her back. She bitched at me for it! She wasn't really hurt. I shattered my sacrum, which is like a nerve conduit coming out of your spinal cord. Oh yeah, that feels great! And then I broke my pelvis in two spots on the right side. And I got like eighteen staples in my back from the windshield. I crawled outta there. And then I started hobbling back . . . and when I put pressure on my right leg is when something shifted and I was just like, 'You're done.' I fell on the ground."

"Damn," I say.

"It starts raining. And the wind is blowing. It's like 40 or 50 degrees. Fucking cold for a T-shirt and shit. I just laid there shivering. Which sucked. Some guy came up a few minutes later, and I guess he called 911."

Two months after that, Huck and his brother-in-law rolled the

1995 Toyota he'd bought to replace his Dodge. He replaced the Toyota with a blue 1995 Chevy half ton with a five-speed manual. That is sitting in his buddy Logan's yard, because Huck is now driving a classic truck: a dilapidated 1968 powder blue Chevy pickup. "There's a trick to it," he tells me the first time I hop in the passenger seat. Then he prattles on for five solid minutes about the many byzantine steps required to start, steer, and stop it.

The other swampers rolled trucks, too. Over the course of the three-day Fourth of July weekend several guys would get banged up and nearly killed in vehicular mash-ups. Smash rolled his brand-new Rambler joy-riding down near Lake Sakakawea. He had to get stitches, and his sister broke her arm. Another swamper called Biff ran his car into a tree in Washington State. His passenger would break his skull but survive; Biff would break his leg, quit work, and move home. The Viking, whom I'd met on my second day of work, flipped his 4-wheeler. He was hot-dogging onto an oil field location on his family's farm (they didn't own the mineral rights) when the front wheel hit a snag and sent him tumbling. He broke his collarbone.

So, while Huck had fashioned himself somewhat self-consciously into a breathing, talking, walking rendition of a prototypical Wild West oil field hand, his behavior was hardly any kind of exception. Huck, in fact, was the rule.

JACK AND DIANE

Climbing into my Chevy at the end of the workday, a rainbow stretches across Diamondback's yard. A storm rolled through on the drive from the work site and now the sun is blazing in the sky, reflecting off the puddles in the gravel. I climb into the driver's seat and turn the ignition.

I found Huck's tales amusing, but stories about car crashes make me uneasy. I've never been a great driver myself. I didn't get my license until I was twenty-three years old, and I really only drove regularly for about six years before I moved to New York, where I rarely used a car. I bought my SUV right before I left for Williston and got in a fender bender before I even made it out of Brooklyn. It was the first time I'd ever driven something so big. For obvious reasons, I was loath to admit this to anyone I worked with, but driving always kind of scared me.

Pulling onto the slick road outside the company's gates, I turn on the radio and I'm met with the opening chords of John Mellencamp's "Jack and Diane." I take a deep breath and turn the radio up. This song always makes me think of Shanon and Donny.

My sister Shanon was a natural leader and a gifted student, president of her school's Future Farmers of America and freshman track star. A brunette with a wide, bright smile, she was constantly compared to Brooke Shields. She was a beauty, the beating heart of our clan, and she took on a great deal of the responsibility in raising me and my siblings. She was a second mother to me. When my older brothers and sister picked on me, Shanon would let me sit in her bedroom while she practiced putting makeup on in front of a mirror. I treasured spending time alone with her. When I was older, I'd learn she also had something of a wild streak. "A real child of the seventies," as Ryan would put it, "she used to sneak out of the house late at night to ride her horse naked across the upper paddock."

My most vivid memory of Shanon was when I accompanied her while she was babysitting a boy in the neighborhood. The boy, about my age, was cutting the heads off all the dandelions in his yard, practicing cruelty, and Shanon saw it was making me upset. "Don't worry," she whispered to me. "Those are seeds. Doing that will make more of them grow." I remember looking into her eyes and both of us giggling. This startling truth was a beautiful secret we shared. I was in love with her. Everyone was. But no one loved her quite like Donny.

Donny was a big factor in my desire to join the army. He was a couple years older than Shanon, big, athletic, and handsome with short blond hair, a wide easy smile, and perfect teeth. He must have been sixteen or seventeen when Shanon brought him to the farm, and to me Donny was an adult and a big kid at the same time. I was five or six years old, and I idolized him. He'd play with Kate and me, racing after us through the front yard, wrestling us onto the living room floor, shouting encouragement as we climbed trees, goading us into snowball fights. He'd sit with us, too, and patiently listen to our gibberish. We hated to see him leave, and we'd follow him out to his car, begging him to stay. But when he took off, he'd hug us, wrapping his big arms around our small bodies and holding us in a rough, playful embrace. We adored him.

Donny was good with parents, too, at least with my parents. He'd talk to my mom and dad about conservative politics. He loved Ronald Reagan and hated Commies. He wore a T-shirt that said, "Join the Army, Travel to Exotic Places Meet Exciting New People, and Kill Them." It was at Donny's goading that I first heard my father talk about his military service. When he was getting ready for bootcamp, Donny sometimes practiced calling funny, bawdy marching cadences in our front yard.

When my mother left my dad, it was Donny who showed up at the farmhouse to pick up Kate, Matthew, Megan, and me. He delivered us to Shanon, who took us to the shelter where we stayed. Later, when my mom moved us to the suburbs, Donny graduated high school and signed up for the army. He sent us a photo of himself in dress uniform seated in front of the flag. My mom had the picture framed and set it at a place of pride on the mantel in our new home. Donny looked like a hero with his square jaw and blue eyes. That spring Shanon won a place on the school's cheerleading squad. The two of them wrote letters back and forth and talked on the phone.

In the spring of 1985, Donny was home on leave to attend the funeral of his grandmother. One rainy night, Shanon accompanied him to a family get-together. On the way home, on a dark, twisty stretch of country road not more than a couple miles from our small house, their car slammed into a tree. Shanon was killed. She was sixteen.

Shanon's death shattered my already broken family. In many ways, she was the glue that kept us together. She was the responsible one, the nurturing one. She brought gentleness out in my father and hardened the resolve of my mother when she most needed it. The wound her death left carved a deep gully of grief that forever spills like an unrelenting river through all of our lives. It remains beyond expression.

Ever since the morning I woke to learn she was gone, grief has been an intrinsic part of my life. At that moment, I was inducted prematurely into a wisdom that shattered my innocence and put me out of step with the rest of the world. Shanon's death, among many other things, was the birth of my solitude, my inability to fit in, my ceaseless searching.

Grief-stricken and heartbroken, left to raise five children without any help, my mom couldn't keep up. After a year of struggling, we moved back in with Dad.

And poor Donny. He woke from his coma, and his first word was my sister's name. When he was told she had been killed, he immediately fell back into shock. He came out of it a week or two later. He would soon visit us on crutches with tears in his bright blue eyes, and we'd tell him we loved him. It never occurred to me to hold him responsible. I do love him.

The accident caused him brain trauma and messed up his back. He was discharged from the army and never finished Ranger school. He'd stick around town, wandering through a series of dead-end jobs and struggling to complete college. He visited us over the years but we'd grown up, and he'd grown more distant.

The last time I saw Donny was after I finished college. It was at a party at my mother's house. Don had received some sort of IT certificate that he hoped would help land him a good job. "I just want to make some fucking money," he said. "I don't care what I have to do for it." I tried to connect with him then, but our lives were on such different paths. I was going to be an actor. At the party, he got too drunk, wandered off, and fell asleep in his car.

"He hasn't forgiven himself," I later said to my mother.

"Maybe you can take him out for a beer," she said, "talk some sense into him."

I meant to, but I never made the call. He shot and killed himself a couple years after that. He was buried in a wooded cemetery in Maryland with military honors.

THE FIFTH OF JULY

The new boarder buys a car and finds a job working at a local body shop. He takes me aside one evening and shows me a series of checks, pulling them out of his wallet one by one, thin and sweaty. He unrolls them slowly and points at the numbers in silence. They are made out to him for $1,000 each. He tells me that he and I should get an apartment together. His conversation is bald but cryptic. He says he is Puerto Rican and from Florida. But he must be on the run, telling fractured tales about ditching a car in Florida, working in Texas, living in Alaska, and dumping off another car on the way to the airport before coming to Williston. I am careful not to ask him questions. I tell him no, I want a place to myself. I know not to invite a vampire into my house.

THE WEEKS PASS in a blur of work and apartment hunting. One night I wake to see the new boarder grunting and talking in his sleep, kicking and punching the air. More nights than not, I abandon the bunk beds and the guttural chorus of snores to sleep in the back of my SUV.

Champ keeps his head down. He lays off the booze, but his presence in the house is as big as it always was. He stays coiled, tense—sober, but ready to spring.

I MAKE AN appointment to look at another room. It is advertised as a basement apartment with its own entrance. The basement is unfinished and the entrance leads down a busted wooden staircase full of gardening tools. I have to duck my head to enter. I walk through a muddy hallway to get to the bedroom. The walls and floor are concrete, and the room is about nine-by-nine with a single dirty window in one corner.

The house's owner is a pale woman in her fifties with a nervous, unhealthy cast. Her fiancé is a small serious man with a mustache. He wears a gold-colored cowboy hat. Their friend, a skinny, blinking woman named Dawn, lives there as well. All three of them stare at me as I stand in the room. The washer and dryer are running loudly. They tell me I won't be allowed to use them, but Dawn says that if I pay her, she will take my clothes to the laundromat.

"FIND A PLACE yet, dude?" Champ asks me.

"No," I say.

"It's hard out there, man. Lemme know when you do. I'm getting calls every day, bro."

"Okay, Champ."

ONE EVENING A couple weeks after the Fourth, I'm sitting in the backyard by myself, and Jesse comes out to join me. He wobbles drunkenly to the table and plops himself into a chair. He is wearing dark wraparound sunglasses in defiance of the slanting North Dakota sun. He'd recently started working the night shift at McDonald's. He said that in the interview, the manager asked him if he even really wanted the job.

"No, not really," he told her.

"Well, at least you're honest," she said. Despite offering signing bonuses, places like McDonald's just couldn't compete with boom wages. This particular branch operated only as a drive-through. The manager hired Jesse on the spot.

Lately, he's been walking to Pepsi every morning at 7:00 a.m., leaving at 2:30 p.m., and walking to McDonald's, where he works until close. He gets to bed around 2:00 a.m. or later, then gets up and does it again. The schedule is grinding on him. His shirt hangs off his thin frame; his face hangs off his skull. But he is behind on rent, Ernie still doesn't have a job, and he has to send money home to his daughter. It doesn't make sense for him to move out of Champ's, he tells me. He needs to live where he can walk to work, and he has to take care of his uncle.

Jesse wipes his face with his hand, revealing a mean purple bruise that swallows his elbow. He took a swing at a Pepsi warehouse coworker, and the coworker knocked his ass onto the concrete. "With my cracked rib, I'm not much of a boxer," Jesse tells me.

The sun is falling down, and it throws dark shadows across the kid's unlined face. He toys with his phone, looking through his sunglasses at a screen saver image of his daughter. He says he is on Paxil, but it hasn't been helping much. He says he takes it because it gets so dark in the winter. I don't have the heart to tell him it is July.

"Was Jesse saying he couldn't wait to get the fuck out of here?

Was he asking you for info on rooms?" Champ asks. I'm putting some food in the fridge, and he's come up behind me.

"No, dude," I say. I slide my bowl of leftovers onto the rack and stand up.

"I was just telling Rickie that everybody shows up and they like us, but then they just fucking hate us by the time they leave," Champ says. I shut the refrigerator door. He has me backed against the wall, and he's blocking my exit out of the narrow kitchen.

"We weren't talking about you, dude," I tell him.

"Okay. Well, whatever," he says. "Just don't fuck me. Okay, dude?"

"I'm not going to fuck you, Champ." His eyes dart around, they can't seem to settle on me, but we stand that way for a long moment before he moves off.

THE BEDROOM IS clean and spacious with hardwood floors and large windows that look out over a tree-lined street. The bed is full-sized and covered in a folksy flowered comforter. The landlord is a large friendly woman who lives downstairs with her husband, a large friendly man. There are four bedrooms and one bathroom on the second floor. She cleans the bathroom twice a week. It's spick-and-span. Downstairs is a microwave and a shared full-sized fridge. There is no kitchen. Rent is $900 a month. I tell the woman I would love to move in. Using the money I borrowed from my East Coast friend, I write her a check. My hand lingers over the number for just a moment. It is the same amount that I paid for a room in the two-story row home I rented in Williamsburg, Brooklyn, supposedly one of the nicest, hippest neighborhoods in the United States.

Two days later, I pay Champ the last of the money I owe him. I don't say goodbye to anybody. I pile my belongings into the Blazer, and I move into a boardinghouse downtown.

Skip and Tammy will move out soon after. Skip gets a job at a truck stop that provides him with company housing. I'll see him when I stop there on my way out into the field. We agree to hang out, and one afternoon I drive him to Minot, where Tammy has moved into subsidized housing. One of the last times we talk, Skip tries to involve me in some kind of internet pyramid scheme selling Christian-themed bathroom products. "No thanks, dude," I tell him.

I text with Gabriel after I leave the flop. I'll even see him once while grabbing gas at the Speedy Stop. We never figure out a time or place to hang out, though, and eventually, we stop texting. Years later, I'll be in Bed-Stuy, Brooklyn, dropping into a bodega with a small hope in my heart that I'll find him, Daniel, or Cholly behind the counter. So far, though, no luck.

Champ and I keep in touch, but I'll never see Rickie again. I'll walk into a laundromat one evening, see the new boarder loading his clothes into a dryer, and turn around and walk right back out. As for Jesse and Ernie, it won't be long before I hear from those two.

RAILHEAD

As the flame-orange sky in the east slowly twists to blue, I follow Bomani up a metal ladder on the side of the first railroad car. Bomani is a small man of indeterminate age with dark black skin and a bright smile. He is one of several Congolese who work at Diamondback Trucking.

The top of the railcar is open and full of pipe, also called joints. The joints are banded together with steel strapping that secures them inside the car. Bomani uses a pair of strapping cutters to snip the banding—they break with barely a sound, then whip violently back from the blade. I snip the next piece of banding using my own cutters. We move across the car, shuffling sort of, so as not to slip on the uneven joints—they're round and slick beneath the rubber soles of our boots.

As we snip the banding, Bomani shakes his head and tells me how much he misses his wife.

"She's a good woman," he says. "My other wives in Africa, they are good, too."

"Oh yeah?" I say.

"Yes," he says definitively.

Soon we have freed all the pipe of banding. Bomani climbs down the ladder on the far side of the railroad car, and I follow. Boots hit the ground for a moment, a couple steps on the tracks, and then onto the next car, climbing up the metal ladder and repeating the task.

Bomani pauses on top of the railcar and motions for me to stop. He removes a glove and, reaching into his pocket, brings out a phone. "I have ten children," he announces, showing me a photo of a woman in a hijab. "I am going to give her many children, too," he says. "Many, many children."

"Uh-huh," I say.

"Yes," he says. "Many wives. Many children. I am King Mufasa! An African king!"

We move across all ten railcars, snipping the banding on each until the sun peeks over the horizon, the orange in retreat.

WITH THE DAY'S first task complete, Bomani positions himself on the first railcar's ladder. A forklift, also called a loader, squares up to the train tracks and dips its forks into the railcar to pick the joints. Bomani guides the operator using hand signals. I stay on the ground, off to the side in the dust.

When a layer of joints is cleared, the loader rattles off down the hill to deposit them onto large metal racks. Bomani climbs into the car and tosses down the boards that separate each layer. I collect the boards and toss them off to the side in a stack to be collected later. The forklift returns, and we do it again.

It's a day in the oil patch with no rigs to move. I'm working at Diamondback's railhead.

Within an hour, a tractor trailer rattles down the access road, and a driver hops out. Friendly, he offers a swamper a cigarette from a pack of Parliaments. One of the loaders peels off and moves down the hill to the metal racks, picks those joints, and moves them to the tractor trailer bed.

When they are done, the swamper and the driver strap down the load, the driver hops back in, and the truck rattles off to Diamondback's pipe yard, where the joints will be inventoried and sent out into the field for drilling operations.

Eventually, I'll also do some work in the pipe yard. There I'll meet a mild-mannered Native American named Phillip with a booming baritone voice and the broad shoulders of an ox.

"It's like a day off for field swampers," Phillip will tell me. "You just gotta follow around the pipe yard swampers and stand there while they work. They're mostly nice guys. Except on Rusty's crew. On Rusty's crew one of them swampers kept giving me shit about being Indian. I told him if he wanted to, I'd step off the yard with him and whip his fucking ass. That shut him the fuck up. But besides that, everybody's pretty nice."

Over the following months, I'll work with Phillip quite a bit and come to really like him. He spoke and worked at his own steady tempo, never rushing and never slowing down; and while he usually came across as very serious, occasionally he would reveal a sharp, deadpan sense of humor that could leave me howling, like when he nicknamed a slow-moving swamper "Walking Eagle."

BOMANI IS AN object of curiosity to the white men at railhead. For many of them, he is the only black man they have ever met. When he says he's never had a cigarette, a swamper hands him a smoke and lights it. Bomani pulls awkwardly on the cigarette and exhales slowly, turning it in his hand, looking at it suspiciously. The swampers and operators sit on boards in an amused half circle. It's lunch break. They watch him, glancing at each other and egging him on. "You like it? What do you think about it? You like it? C'mon, take another drag!"

Bomani takes another hesitant pull. Then another. Each time he looks quizzically at the cigarette and then strangely at the men. They encourage him to smoke more. "It's just a cigarette. Don't waste it!" He inhales again and then falters just slightly.

"I don't feel so good," he says.

"Oh, c'mon. You're fine." They say, "Have some more!"

He sits down on a piece of wood. The men grin and watch him, and he stares back at them. He shakes his head as if trying to clear it.

"Take a dip of Copenhagen," a swamper suggests, and others join in. "It'll make you feel better. Yeah, take some dip; that's what I'd do. It settles your stomach"

The men glance back and forth among each other trying to conceal sly smiles. Bomani appears suspicious, but after a drawn-out conversation, he agrees to try the chew. I sit quietly, a bit off from the others. I'm new, so I'm not expected to join in, but the situation is setting my teeth on edge. But Bomani is playing along, and for the moment, it is harmless. A swamper offers Bomani a can. He takes a pinch of dip, sticks out his tongue, and holds the tobacco over his open mouth as if he is going to drop it right onto his tongue.

"No!" The guys holler. "That's not how you do it. You put it in your lip."

Bomani looks at them in silence until they quiet. Then he moves to do the same thing again.

"Noooo!" they shout. "You put the dip under your fucking lip."

Bomani begins to make the same mistake a third time when,

exasperated, one of the guys moves to him. He takes a pinch of his own dip, shoving it theatrically between his lips and gums. "This is how it's fucking done," he says. Bomani nods solemnly and then follows the instructions.

All eyes are on him. And at first nothing happens. Then Bomani stands up. He is erect for a moment before stumbling slightly. "I'm not feeling well," he says. The white men laugh as Bomani moves in a small, wobbly circle. Finally, he spits the dip onto the ground, a dark wad against the tan dust.

"That stuff is no good," he says pointing at his spit in the dirt. "No good!" The men shake their heads and roll their eyes.

LATER IN THE DAY, I'm alone with Bomani, bored, throwing boards onto a trailer, when he turns to me and lets out a wicked laugh. "Those guys are so stupid!" he says. "They are so stupid! They think I've never smoked a cigarette before? That it makes me act like I'm drunk? They are so dumb. I do it to give them something to talk about. They don't have anything to talk about! They stand around and talk about fucking each other! Up the ass! They don't even talk about fucking girls! I've never heard anything like it."

We enjoy a hearty, conspiratorial laugh.

A FEW DAYS LATER, I am told that on my day off the swampers found a dead bat in one of the railcars. They convinced Bomani to lick it for sixty dollars.

"He wouldn't lick a fucking bat," I say.

But the swamper had filmed it on his phone, and so I watch a video of Bomani holding a dead bat to his tongue for five long seconds. When he takes his hands off, the bat stays connected to his tongue. He stands in the video shaking his hands and wagging his head, until after some effort, the bat falls to the ground.

"What were you thinking, man?" I ask later when I find myself alone with him.

He shakes his head sheepishly. "I'm King Mufasa," he says.

JESSE'S BLUES

J esse and Ernie have a molasses quality about them. They move slowly and deliberately from the grill to the picnic table, from sitting to lying on their backs to standing in the shade. Their faces are lined with grooves of dirt and lack of sleep. Their voices, indeed their entire beings, are slurred.

They carry with them a large aluminum mixing bowl filled with beans and rice mixed together with meat, crusty around the edge. Ernie uses a big spoon to stir it, and occasionally he takes a scoop and brings it up to his face for a chew. They are drinking cheap whiskey out of a brown bag and chasing it down with swallows from a two-liter bottle of flat Squirt.

We stand under a small pavilion with gas grills and picnic tables outside the Stay Well Motel, on an island in the parking lot. It is a perfect North Dakota summer day. The sun is stunning and the breeze ripples through the surrounding construction.

Jesse had called me that morning. Williston had finally whipped him. He was moving back to Georgia. There was a job there, a friend told him, at a local energy company. If he got it, he could make $16 an hour. He would be closer to his girlfriend, and he would be able to see his daughter.

His face is wrapped in sunglasses and a deliriously beatific smile. His shoulders are slumped, his oversized T-shirt billows in the breeze, but he can't help but be amused by his own predicament.

"I got paid on Friday and I left McDonald's feeling pretty good. Having some money in my pocket. I felt like making something happen, you know?"

He walked out the back of the restaurant, eschewing the Million Dollar Way for the access road, strolling up First Avenue, toward Walmart to meet his uncle. They were going to walk back to Champ's place together, to pay Champ rent and go to sleep knowing that for thirty days they had a roof over their heads, and Champ couldn't hassle them about it. He couldn't do them any favors. They'd be granted a reprieve from his bullshit, a little bit of freedom.

"I was walking past the Vegas Motel, and I saw a couple girls standing outside smoking. So, I stopped to talk to them, but they told me to fuck off. 'If you want a hooker,' one of them said, 'go to DK's.' I said, 'Hey now, it ain't gotta be like that.'" Jesse flashes his smile, "I was just sayin' hello."

They let him talk, and he won them over. They invited him into the motel bar to have drinks. Inside, he was quickly introduced to their boyfriends. "Oh, well," he tells me.

Jesse excused himself from the couples' party to use the bathroom. While he was walking toward the back of the bar, someone yelled at him from across the room, "What up, B?!!" It was Frank.

Frank bought Jesse a Jack and Coke. When a bouncer approached him, Jesse said he couldn't find his ID. "This is my tenth Jack and Coke, man," he said. "If you can find my license, I'll give you a hundred bucks!" The bouncer let him stay.

He gave Frank money, and Frank bought him drinks. He gave other guys money, and they bought him drinks, too. He watched

Frank aggravate a group of oil field hands at the bar. Before the lights went up, one of the field hands slapped Frank hard across the face. Jesse left. Uncle Ernie had been calling him; he was still waiting at Walmart.

Just as Jesse stepped out the door, however, a group of young women, dressed in tight skirts and low-cut tops, spilled out of DK's Lounge and started hollering at him. He hollered something back, prompting one of the group, a light-skinned black woman with long fake eyelashes, to strut across the street, walk directly up to him, and grab his dick. She spoke close and soft enough that he could feel her breath on his ear. He followed her back into the Vegas Motel to a shared room.

I imagine Jesse sheepishly smiling, shambling after this woman with a red-faced grin like an elephant led by its trunk. She told him her name was Gorgeous, and she sat him down in a chair in the center of the hotel room and sat on his lap. The room was a mess, he told me, with clothes everywhere and the TV on. She gave him a dance, sliding her ass around, undulating her body and massaging his hard-on through his jeans. Then she got on her knees, took his dick out, and put it in her mouth. Her phone kept ringing, and Uncle Ernie kept calling Jesse so his phone was buzzing, too. Gorgeous pulled down her panties, lifted her skirt, and climbed on top of him. She told him he'd better hurry because someone else needed the room. Her voice grew impatient.

"Hold on, now," said Jesse.

"She was a hooker?" I ask.

"All I know," Jesse says, "is when I looked in my wallet the next morning, I was missing a hundred and forty dollars."

When Jesse and Ernie got back to the flop late that night, Champ had locked all the doors. They tried the front and the back. They knocked and knocked, and no one answered. They tried the windows, but the windows were locked.

Champ had never locked the place up for the entire time that I lived there. In fact, when I first moved into the flop, he made a point of telling me that there were no keys to the house. "I was giving out keys, but nobody ever gave them back," he said. "The door's always open, so you can just walk in."

That night, Jesse and his uncle slept outside. The next day, they got in touch with Champ. He told them he was sorry, he would open up the doors. But when they got to the house that evening, the doors were locked again, and again they couldn't reach Champ.

They spent another night sleeping outside the Stay Well Motel. Jesse curled up in some grass, Ernie in a gated area near a dumpster. By the time I see them, they are defeated. They need to get out of town, and they are all but out of cash.

Ernie has a check for $370 but he doesn't have a bank account, and without a photo ID, he can't turn the check into paper money. All of their belongings, beyond the clothes on their backs, are at Champ's.

And what is it with Champ? If all he wants is money, why doesn't he let Jesse into the house? It doesn't make sense. Maybe Champ is a meth head? Or maybe the power trip is more important than the money?

I tell Jesse I have some things to do. I say I'll check in later that evening. If they can't figure out a place to stay, I tell them they can sleep in the Blazer.

I HEAD TO a small park near the boardinghouse and lie down on the cool grass in a warm patch of sunlight. It is strange having a day off work. I think about calling one of the guys I met when I first landed in town, maybe seeing if Missoula Buck is around. Over the past month, we'd texted several times, but I hadn't seen him since my first week in Williston when I listened to him and the welder's assistant talk about their dads. Maybe he'd want to shoot pool or something. It is mid-July, but the days have been so long that I feel like I've been in Williston a year. I haven't even worked in the patch a full month! Sprawled out in the grassy park, this seems fantastical to me. Then I look up and see Buck sitting on a park bench with a friend. It is as if I have summoned him myself.

Missoula Buck looks good. The night I met him at J Dub's he had a sheepish quality to him. This was the second time he'd "struck out to make something of himself," and the first time he'd washed out of the mines, and wound up homeless. So, there was a desperation to him even on his first night in Williston, an inner leanness his big frame could not disguise. I thought for sure he'd end up working at a grocery store or thumbing his way back to Missoula. But he didn't. He got a job before I did, working as a roustabout for B&G, the same company that had fired Jesse.

Buck is a greenhorn now, like me. He looks big and healthy and confident, imposing even. We joke about our color: dark, burned faces, white around the eyes from safety glasses, and white shiny foreheads. "The oil patch tan!"

Buck and his buddy are whiling the afternoon away, passing a fifth of Fireball back and forth between themselves. They offer me a shot, and out of a sense of camaraderie I accept, wetting my lips with the cinnamon whiskey and then wiping it off with the back of my hand. It is early afternoon. We trade stories. "You been taking too much shit on the job?" I ask.

"One guy was picking on me my first week, running me after

tools that didn't exist, laughing at me for not knowing what things were called," he tells me. "But you know, if you fight in the oil field, guys will just look the other way. I told him if he kept picking on me, we could walk off location and settle it. I'd fight him." Missoula Buck is built like a cannonball. His antagonizer might've been dumb, but he wasn't stupid. "He left me alone after that," Buck tells me.

Buck and his buddy make quick work of their bottle. Shot after shot after shot. I follow them to a nearby bar to play blackjack. Buck's buddy vomits outside the bar and then falls asleep in a chair before staggering home. But Buck holds his own, moving on to tequila and playing cards with a slight wobble and a barely discernible drunken animosity. He's on his way to becoming an oil field hand, alright. I'm tight with my money. I play cards poorly and lose $30. I don't want to be out, or spend money, or get drunk. So I bid Buck farewell and walk a few blocks to a theater, where I sit and watch a movie about giant battling robot aliens. I leave halfway through the film and head back to the boardinghouse.

I don't call Jesse that night, figuring he'll call me if they need a place to crash. I fall into a fitful, guilty sleep.

THE NEXT DAY, I receive my first paycheck as an oil field hand. I make $2,274.58 for 123.5 hours of work over fourteen days. It feels good to hold the check in my hand. It is crisp and clean, and the paper is heavy. From the moment I left Brooklyn, I'd been watching money leak out of my bank account like it was a faulty faucet drip, drip, dripping into oblivion. The drip had gotten me to where I stood, but my resources were depleted. This check would begin to change that. I turn it over several times just to stare at it. I can pay rent. I can make credit card payments. I can begin to pay back my friend who loaned me the money that allowed me to move out of Champ's. If it weren't for that loan, I'd be overdrawn on all my accounts.

Poverty is a habit. My father did not know how to balance a checkbook—he'd admit this to me offhandedly after he retired, when he was attempting to start his own business. My dad was an accountant! All he did, like the men I'd meet in Williston, was work. He didn't have hobbies, or friendships, or obligations within his community. Even so, he couldn't perform a basic function of his job.

When it came to spending, my father never purchased anything without commenting on the cost vocally and silently in an elaborate, embarrassing ritual. Pulling a wad of bills from his pocket, he'd lick his thumb and laboriously peel off the correct amount with the deliberateness of a Kabuki actor enacting a suicide. He did this when he took us to get school supplies. He'd then turn and say, "I just want you kids to know where the money's coming from."

Thanks for the spiral notebook, Pops. His fanaticism regarding our knowledge of "where the money was coming from" peaked when I was in high school, and he wrapped a chain around the refrigerator and padlocked it. The ritual with the wad of bills became the ritual with the ring of keys. I couldn't bear to put myself through it for a snack. For the most part, at that point, I stopped eating.

But now. Holding the check in my hand. Now I could begin to fill the hole I'd dug myself into. On top of that, I could start to rebuild. It was a good feeling. I put the check in my pocket.

When I call Jesse, he tells me they took Uncle Ernie's check to the supermarket, but the cashier wouldn't cash it without a photo ID. They are hoping they have better luck at a liquor store on the other side of town.

I meet them in the parking lot. Both men are terribly hungover with red-rimmed eyes and sallow skin. A heavy hangdog sadness weights their shoulders; they look like a homeless Laurel and Hardy. Before I arrive, the liquor store refuses to cash Ernie's check. He holds the useless paper in a clammy hand, his face twisted in disgust, frustration calcified to jaundice. Jesse is impassive, tired beyond caring. Translucent almost, he looks like the wind could blow through him.

Ernie gives the check to Jesse and Jesse gives the check to me. I ask Ernie to sign the back, and I give him $370 in cash. He immediately offers me a twenty-dollar bill, but I wave it off.

"You're saving our lives," he says.

"It's no trouble," I tell him.

The night before, they returned to the flop ready for trouble, but Champ didn't give them any. He sat in the hot tub cheerful and relaxed, didn't even bother to get out. He said he was sorry for locking the door. They gathered their things and, without incident, left. They tell me Champ wished them the best.

They spent the night in an abandoned bus full of other homeless guys. They were blind drunk, and Ernie nearly got into a fight after stepping on somebody while walking through the bus in the dark.

I drive them back to the motel. With all the windows rolled down, the air swirls around the cab and musses our hair. While Jesse goes inside to retrieve their things—a friend had stashed their belongings in a broom closet—Ernie and I sit in the sun in my Blazer, and Ernie smokes cigarettes. He tells me Jesse has been falling asleep on his feet all day.

"He's just a fucking kid," I tell Ernie.

"He is. He really is." Ernie nods his salami head and draws on his cigarette. He says that he is taking care of his nephew. I know the opposite is true. The kid is saving his uncle's ass. If Jesse didn't feel obligated to take care of Ernie, to find him a place to stay, to vouch for him, to pay his rent, Jesse would have made it. He wouldn't be whipped like he is whipped now. I feel an urge to punch Ernie in the face.

"He doesn't carry himself like one," I say, "but he's just a kid."

OUTSIDE THE BUS STATION on Main Street, a small group of travelers have already gathered. Jesse and his uncle pull their bags out of my SUV and onto the sidewalk. I say goodbye to Ernie. Then I ball up two twenty-dollar bills in my right hand, and I give Jesse a hearty handshake, pushing the bills into his palm and wishing him good luck.

"What's this?" he says, looking into his palm. "Oh, thanks."

I immediately feel ashamed of myself for passing him money, like I'm some kind of big shot, some tough guy, some older brother. Who the fuck am I? I'm tight with guilt. I swallow the feeling down. He needs the money.

Jesse stands listlessly as I climb into the Blazer and pull away. I watch him through the rearview, a stick figure on the sidewalk. If a big breeze came along, he'd probably blow away with it. He is blowing away, of course. With the dust and the dirt on the idiot wind, he's blowing down the road feeling bad—a dumb-kid-leaf, another migrant, discarded. Once again, he'll be crossing damn near the entire country and seeing nothing of it.

I remember the time Jesse cooked an entire ham in the oven at the flop. It was delicious.

"I just haven't had any luck since I got up here," he said to me. I am sorry to see him go.

I'll reach out to Jesse in the months that follow but only hear from him once. The job fell through, and his girlfriend broke up with him, he tells me. He is thinking about returning to Williston. When I eventually leave North Dakota, I try to track him down, but I'll never hear from him again.

THE WILDEBEEST

The badlands, below the sloping hills of prairie, have the features of a fiery mountain range. We drive down into them, descending past the breaks on a ribbon of rust-red road. Cliffs and crags rise sharp around us. The ribbon bends down for several miles and then twists back up toward an isolated tableland. When we crest the hilltop, the Wildebeest shifts gears, the gin truck spins around, and we come to an idling stop, our backs against a steep bluff.

Glaciers and thousands of years of erosion carved the jagged valleys before us. Bluish-gray layers of volcanic ash, brown-orange layers of sand, and black bands of lignite coal wrap like veins across the red-hued scoria that defines the view. We can see for some distance. Everything in front of us is rust colored, more so than the Peterbilt chassis over which we sit, the corroded red of oxidized nuts, bolts, and axles, but not more so, maybe, than the old man's rusty blood.

If I asked you to draw a picture of a trucker, you would draw Wilford "the Wildebeest" Grumlin. His short-cropped hair is white as clouds and his white mustache turns down around his lips. He has clear blue eyes and a pinch of Copenhagen so permanently embedded in his gums that his face seems to have been built around the chewing tobacco. He is in his midfifties but could be a hundred years old.

Sitting in the driver's seat as laid-back as a cowboy sleeping on a pony, he has his Diamondback Trucking–branded ball cap perched on his head in a redneck rendition of a Sinatra-in-Vegas tilt. In all the time I know him, I'll never see him wear a cap or a hard hat straight on his head. The shape of his belly is bowling-ball round, accentuated by his slouch.

He stares droopy eyed into the distance. We are alone and not alone. At the crest of the badlands, we can see a thin line of that golden prairie, but pump jacks and derricks interrupt the view. And in the sonic space between radio songs, the hum of generators swallows the rustle of the breeze.

"This your first day?" he asks me.

"No, but I haven't been doing it long," I tell him. I'm nervous and not nervous. The smartest thing to do, I figure, is to be honest and let some of my nervousness—but not too much—show.

"This is my first day on the gin truck," he says solemnly.

"Well, I guess we'll learn together," I say.

He nods and stares ahead in silence. He's lying. Or it's a joke. Or he is merely enjoying the experienced hand's privileged amusement of fucking with a worm? For whatever reason, what he says is not true. This is not his first day behind the wheel of a gin truck.

An ex-rodeo bronc and bull rider, the Wildebeest was literally born in the oil patch, raised in company housing in the yard of an oil field outfit where his father worked, first in Texas then in North Dakota. Like him, his old man was a gin truck driver. The Beest had been swamping the patch since he was a wild boy of thirteen years. He piloted his first big rig at fifteen, his old man shifting the

gears from the swamp seat as young Beest worked the wheel, clutching down some of these same dirt roads we tear through now, some
thirty-five years later.

We are parked on the new end, the location where the rig will be
built. It is nearly an hour's drive from where the rig is being taken
apart, so we have time on our hands. We sit and wait until we see a
haul truck approaching. It kicks up dirt as it tears through the rust
and the dinosaur bones. When the haul truck crests the hill to the
plateau on which we sit, the old man puts the gin truck in gear, and
we get to work.

A gin truck driver operates mainly in reverse. The Wildebeest
throws his right arm behind my seat, twists his torso, and turns his
neck to look through the latticed steel window behind him. We
tear backward through the scoria toward the haul truck's bed, the
old man's body stretched as straight as an arrow in a bow.

The truck behind us is carrying a shit load. *Shit load* is actually
kind of a technical term. A shit load is a load made up of random
stuff, *a bunch of different shit*. Drivers complain about getting shit
loads because they are more complicated to tie down. More important, because the weight of a shit load is harder to distribute, they
are dangerous to transport. When a trailer is weighted unevenly,
things can get squirrelly.

As a new swamper, I learn to hate loading and off-loading the shit
loads because the rigging is complicated. Unloading a regular load, a
large square box, for instance, it is pretty easy in a glance to know
how to get it off a trailer. But a shit load is like a difficult math problem if it were greasy with invert, dust, dirt, and diesel and worn by
time and weather. And if it were heavy enough to kill you.

In my first months at Diamondback, there was one driver who
always got the shit loads. His name was Bert. He had a big round

face and an awkward caterpillar of a mustache on top of a small red mouth. He was one of the most cheerful, dumb fuckers I've ever met in my life.

"Hey there, slim!" Bert says hopping out of the haul truck and waving.

The Beest backs the gin truck up to the haul truck's trailer so they form a T. I open the swamp side door and drop out of the rig. My boots crunch across the scoria. The Beest stays seated with his head swung back as I climb up onto the trailer.

"How you doing there, buddy!" Bert waves a glove at the Wilde-beest.

The Beest hollers a greeting in return. It is a sound I will hear only in North Dakota, and one I will learn to love and emulate. No one does it with quite the flourish of the Wildebeest. Two words: "Oh yeah," strung out to indeterminate length with barely a consonant to break the wave. "Oooohhaahhhhyeeeaaaahhh," says the Beest.

The driver and I work our way down the trailer undoing the boomers and chains that hold everything in place. The shit load is made up of four weird-shaped steel contraptions. Bolts and metal, all angles, all covered in grease. When the chains are pulled loose and a piece is freed, the Wildebeest, from within the gin truck, operates a lever connected to a winch system that lowers the two poles on the gin truck's bed so they sit over top the metal structure. He then uses a lever to lower the four chains at the end of the truck's poles so I can connect them to the metal to make the pick. The chains and hooks on a gin truck are called *shit chains* and *shit hooks* (I know, I know; more technical talk). My job is to use the shit hooks at the end of the shit chains to remove the shit load.

It sounds easy enough. The difficult thing about unloading any truck is in figuring out where to place the hooks, and whether or not to shorten the chains, and at what length. Keeping the weight level is paramount. An unlevel load can break a chain, cause the load to fall or knock into somebody or, in the worst-case scenario, it can flip a truck or take your head off.

I have just started. I don't know anything, and the decisions I am making can easily lead to the destruction of very expensive equipment or the injury or death of another worker. I am aware of all this. I have to acknowledge it and then try to push it out of my head. Otherwise, I'll be too skittish to do a damn thing. Standing on top of the trailer, I begin connecting the hooks to the metal.

"Not so fast, hot rod."

The Wildebeest stands on the bed of the gin truck and looks down at me. He wears pale blue jeans and a fire-resistant plaid shirt. He stands with most of his weight on one leg, the other leg loose, like a Roman sculpture if it had a fat gut. His thumbs looped in his belt, his hard hat cocked to the side, and his cheek fat with dip.

"Now. Where do you think you need to put that hook?" he asks. His voice has a way of twisting into an incredulous whine at the end of a question.

"Um, I dunno," I say. "Right about here." I try to sound confident as I loop the hook onto the metal.

"Now." He has a way of saying "now" that turns it into a full sentence: "Now why'd you put it right there when you got that little bar sticking out the other side there?"

"Oh, I dunno. Okay. I'll put it there." I move the hook and look back up at the Wildebeest. He looks down at me in disappointed silence. "I mean here," I say, moving the hook again.

"Now. Just take a look at what you're doing," he says to me. "You think that hook ought to go right *there*?"

"Um." I'm sweating beneath my collar. The Wildebeest keeps

looking down at me. Bert stands in the dirt looking up at me. Beneath his cheerful, dull expression I can feel the weight of judgment. "Um, yeah. I think. I mean, I think it should go there."

Clouds roll overhead. Salamanders crawl beneath. The sun sits still in the sky. The earth imperceptibly moves.

"All right," says the Wildebeest. He nods and spits tobacco juice. "Put it there."

Once the shit hooks are properly fastened to the deadly metal, the Beest returns to the driver's seat. I keep hands on the steel as he uses the winch system to raise it off the trailer. He pulls forward, and I hop off the haul truck's bed. I walk with the gin truck, keeping a gloved hand on the steel object to stop it from spinning as we move to the outer edge of location. The goal at this point is to simply get things out of the way until mats are laid down for the rig assembly.

The Wildebeest drops the object into the dirt. I undo the chains, and the gin truck backs up again to the haul truck for the next pick. We repeat the process until the trailer is empty. The driver waves goodbye. "See you later, Slim!" he hollers cheerfully, and heads back to the old end for another load. I hop back into the gin truck to sit swamp side of the Wildebeest. He turns up the radio, and we stare back off into the rust.

As fate would have it, I've met the Wildebeest at a dark crossroad in his life. I don't know it then, but as we sit that day in the badlands, the Beest's wife of twenty years is divorcing him, and his own blood, tainted by a disease no doctor can explain, is killing him. For a man who put family and work above all else, these two assaults on his way of being prove nearly too much for him. Between meetings with lawyers, and appointments with doctors, he'll come to work to be charged with turning the company's most inexperienced swamper into a field hand. And he'll hate me for it.

HUCK FALLS IN LOVE

In the hurly-burly of oil field work, in the dog days of summer, between fistfights and rig ups, Huck falls in love. It was a love that made his atoms quiver, a chemical love, a biological love. He didn't talk to me in detail about it until months later when he felt more comfortable confiding in me, but it informed every decision he made all summer long. He was nauseous with this flu-like crush, this pure teenage love—stinky and raw. It was the kind of love they sang about in the fifties. It defined him as much as, if not more than, his fast driving, drinking, and fighting. It was the reason for the other things, he insisted. At the very least, it was the excuse he leaned on for his other excesses. His hands were sticky with it.

"After that first time I met her it was just like . . ." He struggled to find the right words, then linked his fingers together and pulled hard to show their connection. "There was just no getting away from her, you know. We stayed and hung out till like three o'clock at night all the time."

Her name was Krystal, and she was nineteen years old and "hot as fuck," but she was like the weather, and any change in her affections—a mere tilt of the breeze—could make Huck ecstatic, get him literally jumping with joy, or it could leave him lying slack-jawed on the couch all day. He didn't know how to handle it.

"'Cause she just wants me to be around when Logan isn't around, ya know." Logan was Huck's best friend. And Krystal's boyfriend.

When Logan wasn't working, he and Huck would hang out, get drunk, and get in fights, mostly with each other. The night Logan introduced Huck to Krystal, Huck pounded on him until he had to get stitches. They made up that night, though, with Huck even crashing on Logan's sofa once Logan and Krystal got back from the hospital.

When Logan was working and Huck was not, Huck would drive around with Krystal, and they would cause trouble. Usually dumb stuff, like drinking beers at the movie theater. But it wasn't always so innocent. One night they used a slingshot to bust the windows out of vehicles owned by people they didn't like.

"It was summer!" Huck told me defensively. "It's not like it was cold or anything." Huck was forever defending Krystal's honor, yelling out the window at guys who looked at her wrong, picking a fight with an All-State wrestler who'd offered Krystal a dirt bike in exchange for a blow job. "Expensive blow job," Huck commented. Lucky for Huck, the wrestler shrugged and left.

In what was perhaps Huck's grandest gesture of love for Krystal, he later confided in me that he would take her to Diamondback late at night to drive her around in the gin trucks, teaching her to shift as he clutched the rigs through the dirt yard. He'd be fired if he got caught, and knowing the near-holy esteem in which Huck held big trucks, this was a big deal for him.

However goofy the relationship between Huck, Krystal, and Logan may have seemed, Huck had it *bad* for Krystal. He was genuinely head over heels. But Krystal was trouble. As Huck would say,

"She likes to see me box." And while some young women every-where like to see young guys get in trouble, the stakes for that be-havior in Williston were way higher than they would have been in a normal midwestern town.

I hadn't met Krystal yet, so I only had what Huck told me to guide me regarding her character, and his description was shady at best. In Huck's stories, Krystal comes across as a kind of femme fatale. She's a boy's fantasy, an unattainable, troubled babe.

Occasionally, however, he would give me a glimpse of something deeper that bound the two of them together. And while I found the relationship he described to be infuriatingly immature, I knew that shared trauma created a real bond between them. An overwhelm-ing number of men in Williston struggled mightily with their father wounds, including Huck, including me. But it would be ridiculous to discount the reality of many women who found themselves in the Bakken during the boom. They had their own scars to deal with. Many of those scars were inflicted on them by the beaten boys who'd gathered in town to suck up the grease. But that's the whole cycle, isn't it? Broken boy meets broken girl and makes broken baby. Repeat.

The fact that I never developed any real relationships with women during my time in Williston leaves a big hole in my narra-tive, a song unsung. All I can say is that it felt impossible. I'd ask the guys I worked with, but none of them knew any single women around my age. We all lived in our own individual bubbles. I was working too much to participate in the community. I am not a churchgoer. I rarely went to bars in town for more than a drink or two. About the only time I saw women regularly was when I would sit and write at a local coffee shop. I was in love with the women there. All of them. But I rarely tried to talk to them beyond ordering. I'd heard too many horror stories from guys I worked with about their wives getting hit on while shopping in Walmart. A smile and a hello from me, more often than not, received a dismissive frown in response. I didn't blame the women for their reactions. I looked like every other jerk. I was every other jerk, frankly. I wasn't settling down in Williston. I wasn't looking to start a family there. Town was a wrench I was turning.

So, I can't pretend to know what life was like for women in Williston, not on the level that I understand what it was like for men. I don't know Krystal's story at all. All I know is the way Huck described it, and the effect it had on him. According to Huck, his love for Krystal and his entanglement with her and her boyfriend caused him very big problems. Eventually, it would land him in jail.

WILDER

The tempo would change. The weather would change. The names of the rigs would change. Some of the people would change. But in many ways, every day in the oil field would be the same. The rig moves would become one day of dust and sun, dirt in your mouth, and work. It is August now.

Morning. An alarm goes off at 5:00 a.m. I pull my bones from bed, click the lamp on, and pull on jeans and a T-shirt. I take a moment to massage my hands—they feel crooked as pipe cleaners, swollen, aching. They've been bothering me since I started swinging chains, cramping up, growing rigid, and losing strength. Some mornings I can barely open and close them. So, I soak them in hot water and Epsom salt before sleeping, and when I wake up at night, I run tap water over them—first as hot as I can stand, then as cold as I can stand. Still I wake with my fingers stiff as claws. So, I massage each hand with the other until I feel the blood moving. It has become a morning ritual. When I can open and close a fist without too much discomfort, I grab my coveralls and hard hat and walk out the door.

Cool morning air prickles my skin. There's a sun somewhere, but I can't see it. I can only feel the dew rising. I drive slowly through

Williston's dark web of streets to Diamondback's yard. The circadian rhythms of the city are circular, the streets as busy as ever with trucks and vans, some men heading to work—first shift already stocking their warehouses and hauling their sand and water; some men just now getting off—third shift returning to their trailers and popping that first home-from-work-beer. Talk radio plays softly as I drive.

It is the first day of a new rig move, a common event that fills me with certain terror. If I stop to listen to the echo in my gut or the swarm of bees in my brain, I know I'm a goner. I'll start the day scared, without the confidence I need to push through the toil, the frustration and humiliation. If I start in fear, I've already lost: the day will inevitably beat me. But the fear is there. But I know I can't let it happen. I can't let it whip me.

So, I practice the habits of a good mammal. Miming the moves of a confident man, I walk across the gravel yard in an easy, loping gait and into dispatch. I enter the office, taking a seat behind the glass partition in a dusty rolling office chair. I spread my legs wide and slouch into the chair, pulling down the bill of my cap and barely moving my head. If anyone talks to me—and mostly they ignore me—I keep my responses to a series of grunts, a vague attempt at the language of the unconcerned. If I keep this up, I think, no one will know what a nervous bird I am, no one will see through me. Fake it till you make it.

"You're going out on one fifty-three with the Wildebeest," dispatch tells me, and I nod and go. All the work trucks are numbered, but there are some things my brain has trouble keeping track of, and I'll never remember how to tell any of them apart. So I walk back out into the yard and toward the shed that shelters the trucks. Sharp beams of headlights cut through the velvet dark, throwing light and shadows across the yard, blinding more than revealing. I feel my way toward the trucks. The sounds of motors starting, the stink of diesel, voices of swampers and operators, my boots on gravel. I wander among the rigs and read the numbers on their tailgates until I find 153.

THE WILDEBEEST HAS taught me to inspect the truck before we head out. I walk around it, making sure the tires look full and the bed contains the handful of tools I might need—a swamp bar, a can of grease, slings, and extra shit hooks. Then I climb up into the cab, and he climbs into the driver's seat.

"You get any pussy last night?" he asks me.

"Um, no," I say.

"You dream of pussy last night?"

"I don't think so," I tell him.

"Well, arright," he says disappointed, and he starts the engine.

We ride out toward location, bumping along in the truck, pop country music turned up loud. Silently, the Wildebeest offers me a pinch of dip.

"No thanks," I say.

He grunts. "Suit yourself," he says.

I set my hard hat on the dashboard. "That's a no-no, hot rod," he says. I put the hard hat back on my lap—there is nowhere else to set it. The Wildebeest is slouched in his seat, his ball cap pulled low over truck-stop-bought blue reflective shades, his cheek bulging with dip. We blow through a traffic stop. "Red light, my ass," he drawls.

DOWN THE HIGHWAY, through the White Earth valley and then a left, heading north into the endless patchwork of dirt roads, rumbling over hills and hollows, the kettles sparkling to the left and right—full of ducks—grinding through the mud and then across a wide, dusty stretch until we come to a stop, the first two workers on the new end.

I'M STANDING IN scoria to the side of the truck, I empty my pockets onto the bed—my wallet, keys, and asthma inhaler. I am about to get out of my T-shirt and jeans and slide into my greasers when the Wildebeest stops me.

"You got asthma?" he asks, seeing the inhaler.

"Yeah," I say.

He nods. "What kind of inhaler is that?"

"Albuterol."

He is standing on the bed of the truck behind the driver's side door, the sun finally showing itself behind him—a blaze of red-orange chasing off the purple of night. I don't want to talk about my asthma. I hate to even think about it. There is no feeling more awful than being unable to take a breath. As a child, when my ill-ness was at its most severe, simply walking up and down the steps could result in an asthma attack. In the mornings, I would climb from my bed onto my small mother's back—she's a hundred pounds dripping wet—and she would carry me from my bedroom down the creaky farmhouse stairs to the living room sofa, where I would do my schoolwork, watch soap operas, and read books until bed-time. In the evening, I'd climb from the sofa onto her back again and she would carry me up the steps and deposit me into my bed. I was homeschooled for most of fifth grade, and I continued to miss long stretches of classes throughout middle school. It bugs me to remember how helpless I'd been. On top of that, I'm worried that if dispatch becomes aware of my asthma, they might stop sending me out on jobs. I pull my feet out of my work boots and stand on top of them.

"I first got sick round last Christmas. Didn't feel good, but didn't know what the hell it was," the Wildebeest says, looking past me into the nagging eternity of prairie. "My brother took me to the hospital. The doctor told me that if I'd waited a day longer, I would've died. I knew I was in bad shape, but I didn't think I was in that bad shape."

I pull my jeans down. The hair on my legs rises and goose pimples appear as cool air hits my exposed skin. I wiggle my feet out of my jeans and pull the greasers up my legs.

"What was the matter?" I ask him.

"My blood's all fucked up," he says. "They don't know why, but my blood is attacking my lungs. It's all fucked up." He says it spitefully, with go-fuck-yourself rage. "I'm on all the goddamn medicine. Pills, inhalers, steroids, every fucking thing. It's causing me to swell. They call it moon face. Look at my face. How round it is."

He pushes his neck slightly forward toward me, sticking his chin out. His face, I realize, is almost perfectly round. Like a planet. He looks back at me intently. He has the clearest blue eyes I've ever seen.

I take off my T-shirt and zip the jumpsuit up around me. I am cold now, but it will be hot soon enough.

"My dick is swollen, too," he complains. "Hardly makes me feel like going out and getting any pussy." He turns his back to me and unzips his pants to take a piss, tossing the words back. "If I get a cut, I'm supposed to go to the hospital right away, 'cause it won't stop bleeding. They say even a small cut could bleed me out."

I slide the hard hat onto my head, wrap the shaded safety glasses around my eyes, and take a good look around. The Wildebeest and

I are alone deep in a network of oil field roads, at least an hour's drive from the rest of our work crew and farther still from any town with a hospital. Should any emergency occur, I have no idea how to drive the gin truck. I look up at the Beest, still standing on the truck bed with his back to me. A stream of piss arches in front of him in a thin golden line between me and the sunrise.

He finishes peeing. "Well," he says, "let's walk around and steal some shit."

MOBILE HOUSING UNITS, tanks, trailers, and chunks of the rig litter location. It is unusual for a rig move, but the company in control of drilling operations moved some of its own material before hiring us to complete the job. We stroll around the pieces, investigating the tools. The Wildebeest points at things that could kill me.

"See that cable there?"

"Yeah," I say.

"That snaps. It'll swing back and whack you clean in half."

"Oh."

"Fucking kill you."

"Right."

A few steps later, "See that bucket of grease there?"

"Uh-huh."

"Better not get that on you. Hell, no. Breathe that shit for too long . . ."

"Yeah?"

"Fucking kill ya."

He points at what looks like a winch. "That's called 'the dead man,'" he says. "They use it to lower a tool. It goes miles into the ground. To make sure the drill string is going straight down. I once saw a guy get pulled through there. When they carted him out, he was this big around." He makes a circle with his hands the size of a football.

"Fucking killed him," I say.

"Oooaahhhhyyyeeaaaahh."

AND THEN THE TRUCKS come and the work begins and the sun climbs higher and the dust kicks up and if I wrote the word *dust* over and over for a hundred pages, you'd still have no idea how dusty it gets in North Dakota in the summer. The mud of spring dries, and then it blows.

I was worried when I hired on that I'd have trouble breathing on job sites with all the dust swirling and fossil fuels burning off around me. The ironic truth, however, was that working on rigs, my breathing was the easiest it had ever been in my life. I don't know if that is a testament to the continental climate of big sky country or an indictment of the pollution in Baltimore and New York, but I can count on one hand the number of times I used an asthma inhaler in North Dakota.

I TAKE A LOOK at the metal in front of me. I don't know what it is or what its purpose might be. I have two hooks in my hands. I look for places on the metal object to attach the hooks. I try to guess how the metal might be weighted. It isn't square or symmetrical in any way. Am I overthinking this? I must be.

MY DAD NEVER taught me how to do anything. He didn't know how to do much himself. This was part of my obsession with becoming a good hand. I wanted to become a person who knew how to work, who knew how to accomplish tasks, who could get things done. My father wasn't that guy. He wanted to be. Dad fancied himself some kind of gentleman farmer, but he was incapable of performing even the most basic of chores. Everything he tried was complicated and strange. Nothing ever got done. But every Saturday and Sunday, my father woke, put on an old pair of overalls and a leather cap, and, like a Beckett character, set to a task he would never accomplish.

When the pipes in the farmhouse froze in the winter, the old man would grab a loaf of bread, a knife, and a bicycle inner tube, and he'd force my brother Matthew to crawl under the house with him. I had to help a couple times, too, and I remember shivering on my back in a dirty crawl space full of spiders, as my dad stuffed pieces of white bread into the cracked pipes. He then tied bicycle inner tubes around the holes until the underside of our house became a byzantine network of cracked pipes and torn rubber. "I'm not a plumber," Matthew would later tell me, "but I'm pretty sure that if you look inside a plumber's truck, they aren't full of loaves of bread." Dad never replaced or insulated the pipes, and he never finished the job in the spring, either. It was only in winter, when the pipes were bursting, that he'd force Matthew to spend hours with him under the house on the coldest days of the year.

In the summer, the old man spent his weekends doing one of two tasks. Either he painted the barn roofs or he built fences to keep the horses in. One of the many small tragedies of life on the farm occurred when Dad gave my sister Megan an unbroken Shetland pony mix named Stuffins. Megan was born with hemifacial microsomia, a condition that left her with a disfigured jaw, an asymmetrical face, and a missing right ear. This was never discussed at home, and Megan was advised by my parents to tell kids at school that her ear had been bitten off by a lion in Africa. This was the story she told to her young, bewildered elementary school classmates. As she got older, the lion became a car accident, an idea she got from a boyfriend who lied to his friends on her behalf. When she was teased, Dad showed Megan some boxing moves. She gave a boy in her seventh grade class a black eye and punched and ripped a gold chain off the neck of a boy in high school who taunted her from the back of the bus. When a tall girl threatened her, Dad advised Megan to "get her on the stairs."

The colt my father gave Megan was a mistake that a neighboring farm had been happy to get rid of, and Dad expected Megan to train him. At first, she was excited; she was eight. My father's temper, however, ruled these lessons. He would rage and whip the horse and Megan would plead at him to stop. The pony stayed wild. Since the old man couldn't mount the horse, he rode Megan mercilessly about her supposed inability to tame it.

"Stuffins has jumped the fence," he would roar, slapping his cap onto his head and rushing out the house and up the lane to chase the colt through our neighbors' properties. I remember the amused and horrified looks on their faces as the old man raced after the horse with a lead line and an inexhaustible stream of curse words.

When painting a barn roof, Dad would tie something heavy to the end of a long length of rope and spend the morning attempting to throw the rope over the barn's steeple, building himself into a froth of anger with every missed hurl. When finally the rope cleared the building, he'd haul up a ladder with a bucket of heavy lead paint. He'd spend the day in the hot sun slopping the silver metallic paint across the roof. Occasionally, he'd ask me to play in the vicinity of his work. "In case I fall and bash my goddamn head in," he'd say,

and then snort at the thought of it. He did this for years and somehow never finished the job.

I ATTACH A hook to the metal in front of me.

"No."

The Wildebeest is standing on the back of the truck now. His
hard hat cocked to the side, a thumb looped in his belt.

"No?" I say.

"Uh-uh," he replies.

He keeps staring at me, not offering a suggestion. I remove the
hook. I look at the metal. I take a breath. I move the hook into another position. This has to be it.

"No."

"No?"

"Uh-uh."

"Well," I begin, "where do . . ."

"Where do you think it goes?"

"Well, uh, I thought it went there." I motion to where I'd already
put it.

"It doesn't go there." His voice is high like a whine. The Beest is
grinning now—a mean grin. He spits into the dirt.

"Here?" I say. I put the hook in a third spot.

"Now," the Wildebeest shifts from leg to leg, cocks his head to
the side in disappointment, "why'd you think it'd go there?"

"It's gotta go some fucking place, right?"

"How much do you think that thing weighs?" He sounds like he's
talking to a child.

"I dunno. A lot."

"A lot. That's right. It weighs a lot." He agrees. "Now if you place

those shit hooks on that shit load just any old way, you could flip
my gin truck. You don't wanna do that, do you? You don't wanna
flip my gin truck?"

"No."

"Or cause an accident? Get yourself killed?"

"No, I don't want to get myself killed."

"So where you think you oughta put that shit hook?"

It feels like nothing exists but the burden of those piercing blue
eyes. I look again at the steel contraption we are trying to pick. I
have no fucking idea where to put the hook. I make another guess.

"Well, all right," the Wildebeest says. "There you go."

I'm flooded with a wave of relief.

"Now," he says, "where's the next one go?"

THE DAY IS the same as the previous day and the day before that.
The only change is incremental and personal. The Wildebeest and I
get to know each other:

"You ever fuck a sheep till you wore it out?" he asks.

"Um, no," I say.

"Why'd you stop?"

I DISCOVER ONE way to make time pass: sunflower seeds. I take
a big pinch and place the seeds between my cheek and gums. Then
I suck in hard so they irritate the skin just a bit. The salt tastes
good. I dig the seeds out of my cheek with my tongue, carefully
opening them one at a time to draw out the meat, chew, swallow,
and then spit out the casing. It is a source of true pleasure. I focus
on it intensely, standing in the sun.

. . .

SOMETIMES BOB OLHOUSER, the truck pusher, storms past, a rowdy clamor of man, boots clomping and limbs flapping. Like an oil field Vishnu, he's got a radio, a sledgehammer, a shit hook, and a chain in each of his four magical hands. Banging pins, scaling rigging, sloshing through the dust and mud, the very sky swirling above as if in obedience to his hurled commands. He's howling into the radio—a high-pitched squeal slicing through the bellow of machines—directing the crane, calling the swampers "a bunch of retards," and backing up the bed truck to within an eighth of an inch all seemingly in the same breath. Splattered boot to brow with muck, Olhouser cackles as he goes, gleeful as a murder spree. When he stops in front of me, the earth stops with him.

"I hear you're one of them that likes to work?" he says, looking me over like I'm a tractor he'd consider purchasing.

"Me?" I say, taken off guard. "Who says that?"

"Up in that office there," he tells me, his eyes probing, straining to take me in, to form some kind of picture of me in his big, square head. "They sit around in them chairs up there and they say, 'That new swamper, he likes to work.'"

"They do?" I say, incredulously, but then I recover. "I love to work," I lie. Truthfully, I'd eat glass if I could take a nap. I assume Olhouser's lying as well, but I can't figure out why he would. Later, I'll learn he was speaking truthfully. Dispatch was, in a sense, singing my praises. Of course, plenty of people love to do things they are terrible at, like DJs, so perhaps it wasn't much of a compliment. To me, at the time, I couldn't accept it, but I could tell Olhouser was intrigued by me. In the reflection of myself I see buried beneath his wind-beaten face, I see the possibility of becoming a good hand. It gives me a flicker of hope.

. . .

THE BEEST HAS a small cut on his hand. It's tiny, just a nick. He tapes it up with a Band-Aid, but it keeps bleeding. Dark, rusty blood spills out from under the tape in two thin lines that run around his knuckles. Work has slowed down. "Well, let's go bullshit," he says, pulling the truck parallel to another old-timer's truck. They spend a few minutes grousing about the ineptitude of the other drivers when the old-timer asks the Beest about his bloody hand.

"This goddamn illness," he says. "They can't tell me what it is. They had me in that hospital in Minot. Those bitches stuck me with IV after IV trying to find a vein. You know, to get some medicine in me. My veins kept running away from the needles. Whole arms black and blue and red all up and down both of 'em." He displays his arms, still full of bruises.

"Oh yeaaaah?" the old-timer says.

"Nurses finally found a vein that'd hold the goddamn needle, but then they couldn't stop the blood from flowing out of it. I'd ring the nurse, but those bitches'd never come. I was lying in a pool of my own blood! For. Three. Days."

The old-timer listens and shakes his head. The Beest's face is cold and angry behind his sunglasses. "That goddamn hospital is a goddamn dump. Those nurses down there didn't clean me up."

"Three days?"

He nods his head in disgust. "Three days. Laying in my own goddamn blood."

EVERY DAY IS like multiple days. I step into the shade for a moment, close my eyes, and focus on the breeze. My body aches for rest. I feel close to tears. I'm at the end, and it is barely afternoon. I dream of lying down on the ground, giving myself over completely to the dust. And sleeping. In the middle of location. It would be an act of defiant passivity, a pure surrender, a wild declaration of losing, being beat and giving up. I obsess on the image: myself spread eagle in the sun and the dust, a kind of release, a kind of death, a final joyous farewell to the struggle, and I am comforted by the thought.

But no matter the time of day or the oppressiveness of the sun, the friendly breeze always meets me in the shade, slight but persistent, rolling kindly off the prairie, gently stirring the hairs on the back of my neck, tickling my tanning hide. My eyes closed, I take a breath and the wind rolls over me like water over a dam. I'll go on. I spit out a sunflower seed, take a sip of water, and get back to work.

IT'S SUPPOSED TO be an assisted lift. The mud pump is so heavy that both the haul truck and the gin truck should be tied onto it before it is lowered off the truck's trailer. But the haul truck's bridle—the inch-thick steel cable lowering the tank—slips before the gin truck is attached, and the full weight of the mud pump drops onto the bridle, 140,000 pounds of steel, nearly the same weight as the space shuttle. There's a dull sound like a *thwack!* And the cable strains visibly.

"Get back, dummies!" hollers the Wildebeest. I'm standing at the connection point with another hand, a muscular kid with braces. I catch the Beest's expression and I back away quickly, but the other hand climbs onto the trailer and moves to adjust the bridle, his legs straddling the steel line. "Get the fuck off there, dumb fuck! Get outta there, dummy!" The Beest leans out his truck's window screaming red-faced at the kid. The kid looks at him, then looks at the bridle.

He seems frozen, his expression blank. "Get the fuck off there, god-dammit!" Finally, it clicks for the hand, he shakes his head as if waking from a dream, and he hops down from the trailer, backing far away from it. We watch the taut cable strain from the pressure of the mud tank as it is lowered to the dirt.

"You ever seen a bridle snap back on a guy?" the Wildebeest, still sitting in the driver's seat, a thick arm resting on the window, asks the kid.

The kid shakes his head. He's in the dirt, looking up at the man. "No," he says.

"Well, I have. You get to a point where you get sick of cleaning dead bodies off this motherfucker."

The kid's expression doesn't change. It's a benefit of the hard hat and safety glasses. Nobody reveals nothing. He moves off and the day grinds on.

HOME FROM THE WORKDAY, more than fourteen hours after it started, I pull off my boots and strip out of my clothes. I toss my underwear on the floor of the bathroom and turn on the shower. The stream of water is pins and needles against my burned skin. I turn my face toward it. It empties the lines and crevasses of my face of dust and scoria. I dunk my head beneath it. It untangles the mud clumps in my hair, and the dirt slides off my body into a brown puddle at my feet. I lean against the wall and watch the puddle. It reminds me of a muddy pond in a rainstorm. The kind that brings frogs and tadpoles to the surface. I slide down onto my knees and bend my body over itself like a fetus. I feel the hot water on my naked back. My eyes closed, the hot rain drumming the puddle, the frogs rising, I fall asleep.

DAYS OFF

The sun spills through the windows, splashes over my body. Heat bathes my puffy hands, sore arms, swollen knees, and smarting feet. I wake slowly, keeping perfectly still and staring at the ceiling, lying in bed as long as I can—long after I grow hungry and thirsty and my stomach starts to cramp from holding in my piss. Then I crawl out of bed, pull on a pair of shorts, unlock the bolt lock to my bedroom door, and walk down the hallway to the bathroom.

I drop my bones onto the toilet and, seated, drain my bladder. I am bent over my body, elbows on my thighs, head in my hands, listening to the piss stream into the toilet water. Once relieved, I stand, stagger back to the bedroom, and lock the door behind me.

I lie in bed, in dappled sunlight, in a suspended animation of a sort. I'm on days off. No rigs to move, nowhere to go, and nothing to do but rest. So, I try to rest.

My door remains locked. Thank God for that. I don't want to see them, any of them, whoever they are. I hear them sometimes: the other boarders, and I nod if I pass them at the entrance to the building. "How you doing," we say, not a question. We don't want to know. I don't want to know them. I'm sick of desperate people and hard-luck tales.

The only guy I see with any regularity is a gin truck driver. "Welcome to hell, motherfucker," he says the first time we run into each other.

At the flop, I'd awakened every couple hours, either from noise or light, from snoring, or from sheer survival instincts. The flop never felt safe. But now, in the bucolic quiet of the boardinghouse, with big windows looking out on tree-lined streets, and the still sounds of night—branches swaying in a breeze, the chirp of crickets, an occasional passing car—my body doesn't know what to do. I have to remind myself that I'm in a secure place. In my head I know that, but weeks pass before I get a good night's sleep.

I wake up constantly. Repeatedly, night after night, I leap out of bed and run to the window, half-asleep, disoriented. I am in the "doghouse," the tool shed that sits on top of the derrick floor. It is suspended in the air by a crane. Floating above the streetlights of the city. What am I doing here? They are trying to get me on the radio. Where's the radio? I'm undressed. Where are my FRs? My head is swimming, my heart slamming against my chest. What is going on? I sit and catch my breath. Has dispatch called? No, dispatch hasn't called me. Why aren't I sleeping? Am I allowed to sleep? Yes, yes, I'm allowed. They don't need me yet. But why am I in the doghouse? I don't know. It doesn't matter. Why am I suspended over the streets? It doesn't matter. Go to sleep. Go to sleep, I tell myself. Go to sleep. I return to bed, crawling gingerly back under the covers. Then it happens again. Night after night after night.

In many oil field jobs you work two weeks, then take two weeks off. But not Diamondback. Not for rig moves. I have been working for six weeks now—not long enough for my body to catch up to the changes in my routine. I'm not any stronger, any more resilient than when I started. I'm breaking down. And I never know when I'll have a day off, because I am on call every day. This thought gnaws at me in the mornings, it gnaws at me in the afternoon, it gnaws at me into the evenings. The phone call can come at any time.

"Hey, there," the voice will say, "this is Diamondback. We're gonna have you head out tomorrow in the gin truck with the Wildebeest, okay?"

"Okay," I'll say. I'll say it without hesitation, but I don't want to go out there. Outside of my bolt-locked bedroom door. Into another hot day of dust and dread. I'm afraid. I'm afraid I'll get hurt. I'm afraid I'll be humiliated. Because I *will* be humiliated. I feel like the kid I used to be, the skinny asthmatic in the army jacket, who, after a week of being bedridden by his wheezing lungs, was forced to return to school to face the cruelty of his classmates. I'm afraid I'll never make a hand. But I am scared of losing the job, too. I cannot fail. I couldn't afford to even if I did fail. I'm still short on cash.

I call my brother Matthew in Los Angeles. For most of our lives our relationship had been volatile. As a kid, my brother had a mean streak, an elastic relationship with the truth, and a penchant for drugs and crime. We'd grown somewhat close when we started smoking pot together as teenagers, and we even lived with each other for a year or two in Baltimore, but our alliance was shaky. Matty was a tinderbox—like a 1970s-era Dennis Hopper—when he drank or got high. He lived in such a haze of weed smoke that he was unreliable, and his temper was explosive. He'd later tell me, "I used to smoke pot like other people smoke crack." He could be weirdly charming but also wildly erratic, and I was honestly relieved when he moved from the East Coast to Los Angeles.

But Matthew transformed. While so many of his high school and reformatory school buddies went to jail or overdosed, my brother sobered up in his thirties. He married and had a daughter. The marriage didn't last, but fatherhood did. It became the central pillar of his life. Since then, while our relationship wasn't perfect, he and I had become friends. It helped that we both loved acting. While I was pursuing theater in New York, Matthew was chasing TV work in LA. Over the years, my brother has been run down and tackled by police, shot at and gunned down by criminals on quite a few TV shows. If his life had gone the way it was supposed to, at least one of those things probably would have happened to him in real life.

I tell Matthew about the Wildebeest, about Jesse washing out of town on the Greyhound. I tell him about my busted hands, and the constant, endless, unrelenting stream of bullshit I'm finding myself subjected to.

"It's dragging my dick through the dirt," I tell him.

"You're going to leave," he says.

"As soon as I raise the fucking cash," I tell him.

For the time being, however, I'm in this. Even on days off.

THE LOST PAGES

The Wildebeest tallies up the mileage on the winch truck. On a location just a few hundred yards across, in one day, the truck has gone twenty miles. I've been with it every step of the way. Easily, I've outpaced it, meaning I have probably walked thirty miles. I feel every one of those miles, every one of those steps. My feet are cramping in on themselves. They feel like they're bending toward each other, like I'm somehow growing pigeon-toed. I can feel the strain on my left knee, from where I'd injured it as a kid. And my hands. Still my hands. My hands are twisted up like tough wire bent in weird ways.

"Damn, I feel all busted up," I say.

"Get used to it," he says.

IT IS 99 DEGREES, and there is no relief from the sun, no trees, no shade. My fire-resistant jumpsuit works like an oven, baking me like a cake. I knew North Dakota would be cold in the winter, but I had no idea how hot it would be in the summer. The average high temperature in August is 84 degrees, and it often climbs well into the nineties.

The dust dries out my mouth and tongue. I feel the grit in my teeth. I taste the earth. As the particulate sneaks up my nose, my snot turns to mud. Trying to get it out, I leave a trail of brown-smeared tissues balled up on the floor of the truck's cab. The dirt tangles my hair. It becomes matted with sweat. It forms a crusty border at my hairline. I carry it on my face and clothes in all its forms: dust, dirt, and mud.

I chug water, and I barely piss. I'm sweating everything out. The sweat runs down my back. It trickles off my face in streams, carving ravines through the dirt that clogs my pores. I feel it spill down my ass and the back of my legs. My bones weigh tons. They're heavy, as if they've been replaced by the steel I'm swinging around all day.

Most shit chains are three-eighths of an inch, but the Wildebeest uses half-inch chains. The eighth of an inch difference is excruciating. "Stop your whining about it," the Wildebeest tells me when I don't complain. The truckers complain, everybody who throws the chains complains about their weight, but I don't, and the old man can't stand it. "You're just too weak," he tells me when he sees me struggling.

My hands continue to swell, joints getting puffy, fingers cramping. I can work through the pain, but it doesn't just hurt; my hands lose their strength. My arms are already jelly, and now I can barely grip.

If I wasn't a worm when I walked into the oil patch, I can feel myself becoming one now. I begin to view my body as something outside myself, something detached from my will, a thing I need to compensate for. I am too small for this work. My body is too weak. I try to use my smarts, becoming careful not to grip anything any harder than I absolutely have to. I work to figure ways to throw the chains without fully grasping them, keeping my hands open, using them like hooks. Sitting in the truck between picks, I massage my claws, hot air streaming through the window.

"You gotta start figuring out how to outsmart this iron," the Wildebeest tells me. *I gotta figure out how to get out of here*, I think. *This sucks.*

I arrived in town with under $6,000 in cash and credit, but I had almost double that when I left my job in Manhattan only a couple months before. Buying the Blazer, enjoying an expensive farewell celebration, and crossing the country cost me real money. I have to earn at least as much as I arrived with before I leave Williston or I'll be leaving town whipped. I've been swamping for almost two months now. I sent my friend back east all the money I borrowed from him after I got my second paycheck. Rent at the new place is $900 a month, and I'm learning that my diet of sandwiches, Power bars, and Gatorade from truck stops easily costs me $40 to $50 a day. Because Diamondback primarily employs locals, they offered no per diem or housing allowance for workers, but prices for everything in Williston are still inflated. I have only a couple hundred dollars in my bank account and less than a thousand on credit cards. I can't leave yet. I'm crawling toward my goal.

I AM NOT good at my job. I am always behind a step, flustered, confused, and unsure of myself. When I think I know what to do, I am wrong. When I get it right, I'm moving too slow. My name becomes "Hurry the fuck up!" Exhaustion envelops me. "Hurry the fuck up!" I feel death at my heels, the taunting ghosts of men run down by these merciless machines. The roughnecks last week everyone was talking about? The explosion at the well? How many were killed? How many were injured? I can't remember. "Hurry the fuck up!" the Wildebeest, red-faced, screams.

I'm waving the gin truck around a corner when I turn my back. The shit chains whack me in the back of the head. They send me facefirst into the dirt, my hard hat flying. "You probably don't wanna turn around like that," he hollers out the window. "Now, hurry the fuck up!"

NO ONE GETS fired from the oil field, except for a failed drug test. If you show up and work, as far as dispatch cares, you can keep your job. No one will call you into the office and give you a talking to for bad behavior or laziness. They don't give a shit how bad you are. Dispatch put the bodies in the field and the field can take care of itself. And the field does. Because in the oil patch, on a job location, sixty miles from the office, a different set of rules apply, and they are about as kind and as fair as nature's. If men don't believe you can do a good job, if they think for any reason that your ignorance, laziness, or ineptitude will put them or anyone around them in danger, or even if they just don't like your face—just for the fun of it—they will run you off. They will do everything in their power to make your life such a living hell that you will go home one day and you will not come back. They will replace you with someone else and good fucking riddance. We do not want you, we do not need you. You are a worm, you always were and you always will be. So, go home and worm, you fucking worm.

Two weeks working with the Beest, and he decides to run me off.

"You know," Huck tells me drawing on a cigarette by the mud tanks, "he's asking the other drivers to run over you."

. . .

THE WILDEBEEST KEEPS backing up past the edge of location, ignoring my hand signals, and landing pieces in the mud trench that surrounds the work site like a moat. I have to stomp down into the mud to unhook the metal. My feet make a sucking sound with each step. Muck crawls up my legs, drenching my FRs and weighing me down. Soon I'm caked head to toe in red clay. I stagger behind the vehicle as it peels off for another pick. "Hurry the fuck up!" the old man yells, and dragonflies zip past.

Step after step, mile after mile, I follow the truck. It huffs and snorts like a giant metallic water buffalo. It bests me every step of the way. Eighteen, nineteen, twenty miles. I can't show it's beating me. Twenty-one miles. I can't show it.

"You grow anything on that farm of yours, or you just sit in the house playing video games?"

The old man's comment sticks to my bones. It's too ugly and true. I was the sickly kid, my brothers were healthy. Dad roared at them. He slapped them. He chased them through the yard, pushed them into the dirt and humiliated them. He threw his coffee cup at them. The old man didn't chase after me because I was lying down. He didn't throw anything at me because I never spoke up. My brothers grew defiant, wild, and angry. I became patient, imaginative, quiet, and watchful—a real space cadet. I was the weakling prince, my brothers were the beaten warriors. As an adult, this filled me with a strange shame: my father never hit me.

I also remember full, seemingly endless days scrambling through the woods and climbing trees around the farm. I pulled all kinds of stunts as a kid. In a cluster of trees in our backyard, I once spent an afternoon clambering from one to the other, from oak to evergreen to maple, seven trees in all—the Seven Links my sister Kate and I called them. I leaped from the final tree to the pine needles below triumphant. I climbed a barn roof after a snowstorm and tried to sled off it. I climbed a pine tree in a thunderstorm. Clutching the top branches, sap sticking to my hands, the slender pine bending to powerful, sustained gusts of wind, rain splattering my eyes and soaking my clothes, I watched lightning crack the sky like an electric spider.

I had fake stomachaches until they became real stomachaches. I labored to breathe until I had real asthma attacks. My frailty protected me. My wheezing lungs shielded me from the direct physical results of my father's rage. It wasn't all fake, though. There were ambulance trips and IV drips, pills, inhalers, and injections. Still, illness was a thing I could hide behind.

I shirked from the farm work, hid from my father, and even from the natural beauty of the land around me. I stayed inside, I watched TV, and yes, I spent my days and nights with a video game controller in my hand, commanding a digital avatar who, depending on the game, slayed vampires, outran ogres, or battled spaceships. All while I wheezed in my underwear under a pile of blankets.

I had come so far from the scrawny kid on the sofa, but I couldn't forgive myself for this. What should I have done? Asked my father to punch me?

This sense of shame got to the core of my dysfunction. I'd tried to sing it out and act it out. I'd written about it, raised money, built sets, and advertised ways for people to watch me try to express it. I tried to drink it down. I smoked it and put it up my nose. I tried to fuck my way out of it. I'd used prayer and meditation, mountain hikes, and qigong. I asked Woody Guthrie, I asked Sam Shepard, I asked my mom. But it was still chasing me.

The Wildebeest mocks me, and this shame floods through me. Did I spend my time inside playing video games? Yes, I did.

INSIDE THE TRUCK the radio rattles and clicks with the constant clucking of vulgar hens, a sewing circle of truckers and crane operators nagging at each other in a constant stream of bilious invective. It spills out of the speakers and fills the cab with scab picking and snark. Dressed up as jokes—with all the plausible deniability that provides—the operators compete relentlessly to get under each other's skin. On any given day they may notice a speck of blood on a comrade and choose a new subject on which to direct their collective ire—a new idiot who can't drive straight or should've shut the fuck up, who wormed a hook or had to make more than one attempt at a pick. Woe to the man who lets down his guard.

A common target is Tex. Tex is from Texas, an unforgivable sin, and he is easily riled, a character defect too blatant to ignore. He talks too much, clogging up the radio with his accent and his pride. Tex is easy to pick on. The hens call him "Pubic Hairs." They mock his voice, his home state, and his beard. "What's that sound?" they ask when he talks to the truck pusher. "I hear a sucking sound. Do you know what that is, Tex?" The needling starts early, before sunrise, and it never lets up until one afternoon the Texan is outside his truck, swinging a metal boomer around by its chain, red-faced and screaming to summon the devil. He throws his hard hat—I watch it skid across the scoria—and marches bareheaded toward the truck of his main antagonizer, a sharp-tongued wiseass from Oklahoma. The boomer swinging above his head, glinting in the sun. "Get the fuck out of that fucking truck!" Tex roars at him. "You get the motherfuck outta that fucking truck and I'll take your fucking head off!"

The driver goes pale behind his shades but doesn't move a muscle. All eyes are on him—he wants to flinch—but he holds his ground. Tex stands in the dirt screaming at him: "Get the fuck outta that fucking truck, you piece of shit! You get the motherfuck outta that goddamn . . ." But he can't maintain it. He's got too gentle a heart beneath the bluster. Tex throws the boomer into the dirt. He spits on it, lets out an inarticulate rage-strangled scream, climbs back into his own truck, and pulls off out of location and down the road.

"Wouldn't it be funny if the oil field was run by women?" a lean swamper with the face of a squirrel asks me.

"It is," I say. His forehead twists. He turns toward me.

"No, but like, if everyone was all emotional and on their periods and stuff, like if the oil field were run by women?"

"It is," I say. He stares at me blankly. I exhale. "Actually, I fucking wish it was."

Because as bad as Tex sometimes gets picked on . . .

"Hurry the fuck up!"

. . . I'm getting it worse.

FOR AS LONG as I can remember, I have put my thoughts down on pages. Keeping a diary in a series of notebooks I've had with me since learning to spell. It is a release for me, a therapy, a way to organize the narrative of my life. My journals give me insight when I look back on them. They help me remember. What was I thinking then? What was I feeling? The words I write give guidance to my future self.

When I start swamping for the Wildebeest, my writing dries up. I stop taking notes. I stop putting my experiences on the pages.

When I open my journals, the blank space makes my stomach drop. It is too hard to experience what I am experiencing. It is too difficult to then write it down. I stop writing.

I lie to my friends. The phone calls I make to my pals in New York and Maryland grow shorter or I stop making them altogether.

"I'm doing great!" I say.

"You sound great," they say. And I blather on about the weather, about the work, about the people, about anything but my day-to-day existence. Because every morning it starts anew.

"Good morning, nigger lover," the Beest says as I climb into the truck. "You get any pussy last night?"

He stares at me every time he says it, looking for a crack in my armor, that speck of blood, that hint of fear. And I give it to him. I try to hide it, but I can't. I know I give it to him.

"You see how dirty those mirrors are? The first thing you do for me every morning is you clean my mirrors," the old man tells me, and I climb out of the truck with glass cleaner and a roll of paper towels. I stand on the running boards and carefully wipe down each side view mirror, soaking in the cool morning air and the silence of location before the start of work. Then I get back into the truck and settle into my seat. "Now, does that look clean to you?" he asks, his voice thick with disappointed hostility. I climb out of the truck and clean the windows again. It's a ritual at this point. We play it out most days.

One morning, we run out of glass cleaner and, over the radio, the old man asks another gin truck driver if they have any handy. The other driver is parked on the other side of location, and his radio is off. "Why don't you run over there and see if he's got any," the Wildebeest directs me.

"Okay," I say, and start climbing out of the truck.

"And when I say run," he stops me, "I mean run."

"Fuck off." I drop to the dirt. I begin walking across the scoria.

"When I say run you better run," I hear him holler out the window.

"Sure," I say, breaking into a mock trot, then stopping and continuing to walk.

"I said run, boy!" I turn and see him on the bed of the truck. "You get running, boy, go on, get!"

I raise my middle finger and jog a few paces.

"Go on, boy!" He keeps hollering at me, "You get fucking running, boy! Go on, get!"

And I find my legs moving faster, and I find my breath quickening, and I find my arms swinging as I submit wholly. Like a beaten dog. I run.

LATER THAT DAY, the Wildebeest makes a mistake. I'm standing on a mat, signaling him to back up when his elbow knocks into the winch control inside the truck. The chains, all bound together at the end of the poles, drop suddenly. I hear a *whoosh* as they hurtle to the ground inches from my head. They smack into the mats with a clanking thud. They could have broken my neck. I don't say any words to the Beest about it, but I am sure to hold his eyes for a moment. You're losing a step, old man, and you almost cost me my head.

AT THE END of the week, the Wildebeest and I tally up our time. We have worked the past fourteen days straight, 95.5 hours in one week, 172 hours in two. I try to let the numbers sink in, but they wobble at the edge of my consciousness—*95 hours of work* in one week. I feel the figure seesaw: *172 hours of work* in fourteen days. I didn't even know that many hours existed. And I feel the numbers fall, tumble out of my brain, down my raw throat, into my empty belly, and down to the floorboards beneath my swollen feet. I'm sitting in the swamp seat like a marionette with its strings snipped, a tired, aching shell of a thing: I've got a bruise for a torso, spaghetti for limbs, and broken claws for feet and hands. And I've got dirt. I've got dirt in the creases of my fingers. I've got dirt in the lines of my face. I've got dirt under my balls and down the crack of my ass, in my nostrils, in my teeth, in the crow's feet at my eyes. Dirt in my tried and tired soul. True grit, I guess. If I weren't encased in it, I'd probably collapse.

The old man is lecturing me on rigging, his maw a senseless, wet noisemaker. He yammers on and on, a blathering display of truckermouth diarrhea. I don't listen. Because I don't care how Tex throws his bridle, how many tons a mud tank weighs, or if the cable on the haul truck was strung up wrong. I don't care that that dumb fucking worm from Minot got promoted to driller for sucking the company man's dick in the tool shack. I don't give a damn about any of it.

Outside the window, the prairie in all its empty sameness, in all its boundless anti-glory, passes by. Like the old man's droning voice, it doesn't start or end. I'm so tired I could piss all over it.

But something happens.

"I'm getting better," I hear myself interrupt him loudly. Too loudly. My voice slices through the old man's monologue, stops him midsentence.

"What?" he says.

"I'm getting better." It's my voice again. I barely recognize it. "At the job," I say. "I'm figuring it out."

The old man grimaces. He shifts uncomfortably in his seat. "You gotta . . . ," he starts.

"No," my voice stops him. Coming from somewhere else, confident and desperate at the same time—a needy bellow, booming like a pip-squeak. I can't quiet it. "From when I started. To now," I say, "I'm learning. I'm getting better at the job. I'm making a hand."

We pull into Diamondback's yard, and the old man parks the truck in slow, ragged silence. He turns and looks at me, reckons me. His arms slung over the steering wheel. He stares at me placidly. He doesn't need to tell me how long he's been doing this. He doesn't need to say how many miles he's logged, how many hours he's hauled, how many rigs he's moved across these desolate plains, and he definitely doesn't need to count for me the number of swampers he's whipped in the meantime. Whipped and broken.

He puts it to me simply: "I could kill you on location. If you got tangled up behind the truck somehow. Nobody would think anything of it. Just a work-related accident. Nobody would blame me."

He gazes at me an interminable moment. I don't flinch, and I don't look away. But I feel something change, something unidentifiable, deep inside of me, something tender and gentle, something small and sacred. I feel it quiver and then harden.

The Beest had seen me struggling in the sun one day in my jumpsuit, and he brought me a couple lightweight fire-resistant shirts. They were from a previous job of his, and his first name was embroidered over the front left pocket. When I ordered food at the truck stop, the women behind the counter read his name off my shirt and started calling me Wilford. It was disorienting at first, then I felt too embarrassed to explain myself after the ladies had done it once or twice. It also made me feel nakedly intimate with the Wildebeest. He'd just given me a couple shirts, sure, but he'd also somehow given me his name.

My father had threatened to kill me, my mother, and all my siblings. And like a tree growing on the side of a rugged mountain, I'd conformed to this certain brutality in order to survive. The dynamic I had with the Beest made sense to me, not in my head, but in my blood and my bones. It was woven as if into my genetic code, passed down to me in the trauma passed down to my father from his father and his father before him. This experience fit my understanding of life in a fundamental way.

But is this why I came to Williston? Is this the true reason for my being here? Had I really just given up everything I'd built in my life and driven 2,000 miles to this desolate patch of land so I could find a man to abuse me?

I watch the old man as he slowly starts to move. He picks his trash up off the floor of the cab—water bottles and food wrappers. He opens the door, turns his back to me, and climbs out of the truck. I lean back in the seat a moment longer. I don't shut my eyes. There's a new gun in the house.

JUST FINE

Brooklyn, for me, in 2012, had been the land of milk and honey. I was living a solid middle-class existence for the first time in my life. I could pay my bills on time. I had credit cards, a slick apartment, and nice clothes. I'd go out to eat, and drink fancy cocktails with orange rind garnishes. I had weed delivered to my door and I'd talk over the finer points of sativas and indicas with my delivery guy like a couple of pothead sommeliers. I bought things—once spending $500 on an Xbox and a bicycle in the same day. Thirteen-year-old me slapped high five with thirty-five-year-old me. I got stoned, raced around, popped some wheelies, and then killed a thousand motherfuckers playing *Grand Theft Auto* while smoking joints all night long. I wasn't rich, but I was content. I was doing *just fine.*

Just fine can be hard. I wouldn't fully realize it until years later, but I didn't know how to be okay. The stress of the gun in the house I'd grown up in had nagged my short arms and legs to length, had pulled me taller. Through grade school, high school, college, and after, it molded me into the shamble of a man I'd become thirty years later.

After the death of my sister, when my family moved back into the farmhouse with my dad, we experienced a period of relative calm. Mom began fixing up the old house, scrubbing and painting the walls, mending furniture, and hanging plants. Dad was happy to have us back. He'd been humbled by the separation and seemed ready to turn over a new leaf. But it didn't last long. The old man simmered with resentment. He began dividing time into two eras, an indelible Eden called "Before Your Mother Took You Kids" and a hideous fall from grace known as "After Your Mother Turned You Kids against Me."

Dad could be so over the top that he was accidentally hilarious. I'll never forget the old man seated at the kitchen table one evening eating his dinner alone, using a sock for a napkin, Phil Collins's "I Don't Care Anymore" playing on a small radio, Dad singing along with his own words, "I don't give a goddamn anymore! I don't give a good goddamn!!" You get used to things like that as a kid, though. Time passes, and when your father begins actively blaming your brother Ryan for the death of your sister, you just put your books in your backpack with your bagged lunch and you trundle off to school.

Then something happens that is too terrible to accept or ignore. I was in high school when my younger sister, Kate, told me Dad had sexually abused her. I told her I knew but Kate told our mother, and Mom spoke to a therapist. The therapist said, "I have to call the police." My mother had twenty-four hours to confront my father before the cops arrived. Kate remembers that she and I stayed upstairs in her bedroom while Mom talked to Dad downstairs. We barricaded the door with a dresser and sat by the open window holding hands, ready to climb out onto the roof and make a run for it. But my father did not explode. He listened to my mother and walked up to his bedroom.

Later that night, Kate and I huddled in the kitchen with Mom to find out what had happened. When the old man walked down the steps, we scattered. I dove into a closet and hid under a pile of jackets. Then I watched through the cracked closet door as my father entered the living room, punched a framed photograph sitting on an end table, and sat down on the couch, staring straight ahead, his face ashen, as etched in rage as a statue of a furious pre-Christian god.

I don't remember how the evening ended, or the week, or the month, but the police did immediately not make my father leave. For some time after that night, Kate, Mom, and I were stuck sharing the farmhouse with Dad. Ryan and Matthew had left home. Megan, after showing up drunk to work and getting fired from her job waiting tables, fled to the beach to party with a friend named Chastity. It was summer. The overgrown blackberry bushes and trees of heaven pushed up against the long, lean driveway—the only way out of our isolated valley—threatening to swallow it. I honestly don't remember much from that period of time. Kate recalls an era of constant terror.

Eventually, we fled and found ourselves again living in a shelter. We gave statements to the police, they drew up a restraining order, and the old man's worst fear came true: he was banished from his land. We moved back into the farmhouse but continued to live under the lingering threat that he would show up one day and finally accomplish what he'd talked about doing for so long: he'd kill us all.

When I graduated high school, my father offered to take me out to lunch and, hesitantly, I agreed. I thought he might give me some money. On the way to the restaurant, he took a long detour, driving us down the thin, winding road where my sister Shanon had been killed. He pulled off the asphalt and onto the dirt, parking in front of the tree Shanon's car had hit. "Well, here it is," he said. For several minutes, the two of us sat silently, staring at the tree, the very place my eldest sister's brief life had been yanked away from her. Shanon had never graduated high school, I realized. I was two years older than she would ever be. Abruptly, Dad put his car back in gear and drove us to Olive Garden. At lunch, he handed me a check for three hundred dollars.

Months later, I told him I didn't want anything to do with him. It was hard, not because of any affection I felt toward him, but because he scared the hell out of me. Not long after, however, I was arrested and slapped with a federal drug charge after getting tackled by cops while smoking a joint in a park. I suddenly faced not only jail time but also the real possibility of losing all the federal loans and scholarships that were allowing me to enter college, the only way I saw of getting out of the farmhouse. Despite the fact that my mom was generally a very nurturing presence in my life, she was a hard-ass when it came to law and order. I knew she would tell the police to keep me in jail. So I called Dad, and he bailed me out.

For years, a very uneasy alliance would continue between my father and me, off and on for a decade, until as an adult, Kate reconciled with him. They actually became good friends. He gave her away at her wedding. It was a remarkable thing. I began to invite him to my plays and music gigs, and most of my friends got a kick out of him. They thought he was a good guy. He could be charming in an odd way. Blunt, and awkward, he liked to whistle when he walked around, and he'd chat up strangers everywhere he went. He was game. He liked to dance. He had a big, barking laugh that could absolutely consume him. And he understood adults better than children. He didn't know how to be a father, but he knew in some ways how to be a pal. To most folks, I'd guess he seemed a little eccentric, but amiable.

His good mood wouldn't last. The last time I'd spoken to him was only a couple years before I moved to Williston. It was my lunch break, he was on the phone, and I was strolling through Midtown on my way to grab a sandwich. I don't remember what got him started, but after years of letting sleeping dogs lie, paranoia again began to consume him. A few days earlier, Kate had given me a heads-up. "He asked me to grab a coffee with him," she told me, "and I could see it right away. The cockiness. The swagger. The old dad is back." Over the phone to me, he again threatened my mother. He again referenced his gun. I hung up and had the first full-blown asthma attack I'd experienced in years. My parents had been divorced for nearly two decades. The farm was long gone, but Dad still knew where Mom lived. Would he actually go through with it this time? It was impossible to say. When I recovered, I called Mom and told her not to answer the door. If Dad showed up, she should call the police.

That weekend, over whiskey at a rock-and-roll bar, I drunkenly told my friend Veeka about my father's threats. Veeka was Brooklyn-bartender tough, voluptuous and tattooed, with big eyes and a wicked smile framed by black bangs. She had an edge to her, but when she listened to my words all hardness fell away. "That's not normal," she said to me softly. "You know that, right?" Her leather jacket creaked when she put a hand on my shoulder to make sure I heard her. But I didn't fully understand. This *was* normal, goddammit.

Over the past few years in New York, in the period that should have been the happiest and most carefree days of my life, I'd developed some bad habits. My hunger for the dual thrills of drugs and sex had grown ravenous. One afternoon, months before my father's phone call, I found myself in a brothel with the cocaine sweats after smoking crack, telling a hooker about the recent tragic death of an old friend. This felt good. It didn't feel bad. It felt normal. I was okay with the whole scene. *This is it*, I thought. This is where it's at. *This* is just fine.

I'd blow lines before meeting internet dates. If they didn't want to have sex, I'd make a phone call and drop a few hundred dollars to meet a woman in a hotel room who did. It wasn't the sex or the drugs that appealed to me most. It was the stress. The feeling *before* the action is what I craved.

I got my coke from a group of Polish dudes who drove black SUVs around Bushwick. I'd text them and then meet them on a random corner. It was always as I opened the door to their SUV that I felt my best, my most primed, my most animal, my most aware, tottering on an edge I relished. I'd slide into the car and make the deal, hopping out a few blocks later.

The excitement I got paying for sex was similar. The feeling I craved wasn't just sex—that was an added bonus—it was the phone call to the woman, setting up the time and location, then the walk to the hotel, calling from the lobby for the room number, then the elevator to the room, my headphones in, Warren Zevon howling "Werewolves of London" into my ears, wondering what would meet me on the other side of the door. Would it be the woman in the picture? Or would she be older and heavier? Would there be a man with a gun? Would he be a cop? My heart would bang through my chest as I stepped into the room, trying to take in the layout, blinking my eyes to adjust to the light as quickly as possible. Once I found the situation as advertised, the next steps were pro forma. I'd get laid, then leave and go get fucked-up on whiskey.

I didn't realize I was an adrenaline freak. I didn't understand that I'd grown so used to discomfort as a child that, as an adult, I actually found it comforting. And I don't blame my father for anything I've done as a grown man. I own all of it. But it wasn't until later that I started to unwrap any of this stuff. At the time, I just knew I liked exciting things. I wasn't self-aware enough to see that when stress wasn't there, like a junkie looking for a fix, I went out and grabbed it. Even when things are good, you can put a gun in your house. Even when things are *just fine*.

Then one day a powerful, sustained blast of wind rolled over me, plastering my clothes to my body's frame, flattening my hair. The hurricane hit.

THE ROCKAWAYS

The night before Superstorm Sandy smashed into New York City, I saw gas lines twisting around the block in my Brooklyn neighborhood like vines choking out a tree. The usual frenzied snake of traffic had coiled into a metastasized knot. Some people stood in line with their cars, arms crossed, red plastic gas cans in hand, shuffling their weight from foot to foot. *How can it be*, I thought, *that New York can't get gasoline?*

Sandy would prove to be the most destructive hurricane of the 2012 season, making landfall in Jamaica, walloping Cuba, tearing through the Caribbean, and then roaring her way up the East Coast of the United States. In New York, Lower Manhattan, Red Hook, and the Rockaways bore the brunt of the storm's force.

The Rockaway Peninsula, on the edge of Queens, was one of my favorite places on the planet. During many a summer weekend, I'd scurry out of my apartment and through the bustling Brooklyn streets with a swimsuit and a cooler, climbing down onto the dank platform of the L train and transporting myself an hour later onto a beautiful thirteen-mile stretch of sandy Atlantic shoreline—seagulls swirling above, a joyful sun shining down.

The Rockaways are the epicenter of the American melting pot in the same way Manhattan is. But better in some ways, more unexpected. Stripped of suits and skirts, watches and shoes, and shining with oil and sunscreen, from Breezy Point on down, the beach teems with barely-clad bodies that span the spectrum from black and brown to yellow and pink, from short to tall, and from bony to brawny in a glorious visual manifestation of Whitman's America. In 2012, real estate developers hadn't yet discovered it. I'd spent many a lazy summer afternoon there with sand in my hair, salt drying on my reddening skin, beers in the cooler, and friends by my side.

But summer had ended. I hadn't been to the Rockaways in over a month when, on the night of October twenty-ninth, with a full moon in the sky and the Atlantic at high tide, Superstorm Sandy slapped the peninsula in the face.

A few days later I ran into a fitness instructor I'd briefly dated, and she invited me to join her group of friends for a day of volunteer work on the peninsula. With no gas stations operating, transportation was at a premium, and these folks had a van. I met them at 7:00 a.m. the next day, and we headed into the heart of the destruction.

Trash was strewn everywhere, scattered across the roads and captured in the branches of trees like apocalyptic leaves. Sand from the shoreline covered the streets. We came upon a sailboat capsized in the middle of the ragged road—its hull smashed, its mast snapped in two—and had to back up and find a different route to get around it. Flooded-out cars sat abandoned on deflated tires with their windows shattered, the doors left opened. Some buildings had burned down. The blackened shells of their facades lined the road like singed and broken tombstones. Like everyone I knew, I'd read countless articles about the results of man-made global warming, but this felt like the first time I was seeing it up close.

It was early morning when we arrived, and quiet. We climbed out of the van, grabbed respirators, and walked into the closest residential neighborhood. We hauled mattresses and furniture out of basements, living rooms, bedrooms, and kitchens. We carried personal things: photos and paintings, trophies, and knickknacks onto the street, tossing them into piles at the end of driveways, turning waterlogged junk into monuments to the storm's destructive powers. The people we helped were tired, many of them still in shock at the devastation that had been visited upon their homes. Some seemed grateful, but they were hardly effusive. We were helping them throw away their lives.

Over the following months, I spent all my free time in the Rockaways. Mostly, I organized small crews to do demolition on flooded houses. Thousands of homes had been filled with raw sewage during the storm. Their interiors were circled by a thin brown line about three feet off the floor. With respirators, sledgehammers, and crowbars, we moved from house to house that fall and into the winter, banging out drywall.

The work woke something up inside of me, something that had been dormant for some time, something simple and elemental. I liked being tired at the end of the day. I liked having sore arms and sore feet. I liked getting dirty. On top of that, I loved the opportunity to do something I felt had meaning. It woke up my hands and my heart.

Where I was felt like the center of the world. People were emotional, amped up, angry, and frustrated but funny and caring, too. One of my favorite places on the planet had been devastated, but the opportunity to help rebuild it felt like a gift. It was a sad time, but also joyous.

For about six months before Sandy hit, I'd been obsessed with news of North Dakota's oil boom. It was a joke I had with myself, a flight of fancy, something I would never actually do. I was too scared. But after weeks of pounding out walls, the weather turned and most of the relief effort dried up. I ran into an old friend at a Christmas party. "I'm going to North Dakota to work on oil rigs," I told him impulsively, not sure how much I believed it myself.

"Oh yeah?" he said. "That sounds like something you'd do."

The following week, I put in my notice at work.

GLENDIVE

Hazy glamour shots of former Miss Montanas line the walls of the motel lobby, flanked by taxidermy: stuffed deer, elk, and pheasant. I'm in Glendive, Montana, a town of five thousand, incorporated in 1902 as a stop on the Northern Pacific Railroad line, set between the badlands of Makoshika State Park and the Yellowstone River. In 2015, a pipeline break will dump 31,000 gallons of Bakken crude into the Yellowstone at Glendive, poisoning the city's water supply. Montana's Department of Environmental Quality will collect a million-dollar fine from the company responsible, which just happens to be Diamondback Trucking's corporate parent.

It is Labor Day weekend, and I learn there is no holiday pay or per diem for out-of-town work. I wasn't even told I'd be staying in Montana. The dispatcher just said, "Pack a bag," before hanging up on me. When I check into the motel in Glendive, I find out that I'm sharing my room with the Wildebeest. What fun. We toss our luggage in the room, clean up, and head to the motel bar.

I order the elk burger and a beer. The old man gets a steak and a glass of Pendleton Whisky. We eat in a rambling, bored silence.

Country music plays softly, and the bartenders don't pay us much mind until the Wildebeest gets one of them talking, a middle-aged woman in an oversized T-shirt with the image of a wolf on it. At first, she is warm and friendly.

"Where y'all from?" she asks.

"Williston," the Beest says.

"Oh," she responds, cooling. "I don't go much up around that way anymore. What kind of work y'all do?"

"Oil field," he says.

"Everybody's got to do something, I guess," she says, cold now, and moves down the bar.

The Beest turns and looks at me. He's got no dip in his lip, but he still holds his jaw like he does. The blue in his eyes comes to the surface for a moment, a passing sadness. When it is gone, it is gone. He puts it away like he's locking a painting in a closet. Then his eyes flash. Anger, that old standby. He starts railing against a truck driver he claims whines too much.

"I can't get over how much you fuckers complain," I tell him.

"It's a thing," he says. "I used to be a pretty happy-go-lucky guy before I got behind the wheel of a rig."

"Oh yeah?" I goad him, "Where'd it all go wrong, Beest?"

"As soon as I set foot in this goddamn oil patch." He says it so definitively that it stops the conversation. He first set foot in the oil patch when he was twelve years old. He picks up his glass of whiskey and takes a silent draw. I turn away from him and swallow my beer. The seconds tick by.

I ask him about the rodeo. "Tell me about your bronc riding days," I say. He removes his dentures to show me where a horse had kicked him in the face. Then he starts talking oil field accidents. The stories begin woefully, in a grim voice with a set jaw. They come from darkness, but soon they buoy him, and pride lightens his tone. Like war stories, oil field stories stand as testament to the survival of the teller. The more gruesome the account, the tougher the sonovabitch who lived to tell the tale. It's a cruddy, cruel bravado.

"One guy got hit so hard on the side of his head that his face moved around to the other side of his head. His whole goddamn face," the Wildebeest says.

He tells these stories on the job site supposedly to make me smarter and more careful, but also to mess with me, see if he can scare me, to kill time, and to boast. Here at the bar he can't maintain it, though. He exhales and swirls his drink, suddenly deflated.

"These goddamn nightmares," he tells me staring into the whiskey. "I dream I hurt somebody driving around location. I'm driving the truck and someone gets hurt. Real bad. And it's my fault."

His mustache droops nearly into his drink. The drugs he takes for his blood make him puffy like he's hollow inside or filled with marshmallows. The emptiness of his life made manifest by a combination of steroids and antibiotics, he looks like a potato sack set on the bar stool. I am faced with the shocking contrast between the hollering, shit-kicking, ass-whooping operator of the gin truck and the sad, lonesome man beside me, a picture of profound regret.

I see the Beest in that moment as an aging gunslinger, a killer who has lost a step. He's old. He's sentimental. He's going to slip, and I don't want to be around for it. Because there's an element to this nightmare that gives me pause. The guy he hurts in this dream? Real bad? If it were to come true? That would be his swamper. That would be me. And I don't want to be behind the truck or under the load when it happens. I don't want this sad old man to make a mistake and kill me.

. . .

AND THEN THERE we are again. Perched on a hillside, sitting in a truck, staring down a long, winding dirt road. The new end. Haul trucks coming from over a hundred miles away, we wait for them, watching the day crawl slowly under the sun, shadows stretching long across the dirt.

The Wildebeest: Do you want a cheese curl?

Me: No, thanks.

The Wildebeest: It'll stop up your butthole.

Me: Yeah, no thanks.

More silence. The wind rustles the switchgrass. I stretch and yawn, resigned to the tedium and terrible company. Then Bobby Lee shows up in his work van with a couple swampers. I see the outline of his cowboy hat through the windshield. The sheriff's in town. Smash and his operator, Charley, pull in with a crane and a pickup truck, Tex and Porkchop arrive in the flatbed. The morning's silence is swallowed up by afternoon's progress. The haul trucks arrive, and we get moving.

It is a good workday, and I get busted all to hell. Tying onto a load, a hook swings out and hits me hard in the back of my hand, swelling it up immediately. I later bang my pinkie against a boomer. I lose all feeling in it for a couple minutes. When the feeling comes back, man, it smarts. Midafternoon, I trip over a bridle and fall face forward into the scoria. I'm on my hands and knees in the dirt. "The work is up here," the tool pusher comments, dry as dust.

That morning, on the way to the lease, we spied a herd of antelope galloping across the plain. On the way back to the motel, at the end of the day, we stop for a deer, a doe, standing in the middle of the road, unfazed by the truck. It isn't until a coyote scampers around on the driver's side that the doe bounds off. I push the truck past fifty miles an hour down a gravel dirt road, kicking up a comet trail of dust that billows off across the hill. It tumbles over and over and over itself like it's gonna wrap all the way around the world.

That night, all the hands hit the motel bar. We make for a rowdy bunch, grabbing beers and burgers and filling up the tables in our scruffy work clothes. Bobby Lee and a haul truck driver take seats in a booth with some middle-aged ladies. At one point, I hear one of the women say the words "solar energy," and I almost spit out my beer.

The Wildebeest and a haul trucker with a "Married in Sturgis" tattoo on his forearm ask our waitress for a plastic bag. When she brings it over, the Beest puts it on her hand and starts massaging her arm. She gives him a dubious look. He says, "Do you know what to do now?"

"I don't," she says.

"Well," he says, "I just wanted to find out if you knew what to do in the sack."

She laughs good-naturedly, takes the bag off her hand, blows it up, and ties it off so it is full like a balloon. She gives it to the Wildebeest. "Now you can tell all your buddies I gave you a blow job," she says, and saunters off.

The haul truck driver gets blitzed. He sits with the Beest and talks about the boom in the eighties. "You'd walk right into a bar, see a table full of roughnecks, drop a hundred-dollar bill on the bar, say, 'Get these boys a round,' and then go take a piss."

I am about two months into the job, and this is my first chance to put myself on equal ground with the other men. I still feel timid. I still feel like a worm. I am still learning in the patch. Here at the bar, though, the opportunities are different. We're walking the same floorboards and drinking the same booze. I've worked as a guitar player my entire adult life. I can hold my liquor. I look around the room full of hands. I can go toe-to-toe with anyone here.

"You weren't any good today sober, Mike," Bobby Lee says, confronting me at the bar in his soft voice. He's wearing his usual blue denim western-cut shirt, blue jeans, and cowboy boots, with his ratty cowboy hat pushed back on his head. "We better not let you drink any more tonight."

I laugh it off and sip my beer, but Bobby repeats it to the other hands, casually, like it's just plain good advice. "Don't let Mike have any booze. He's no good sober, can't imagine what he'll be like hungover."

"Go fuck yourself, Bobby Lee," I tell him, hardly a stinging rebuke but at least a confident one. We exchange a few barbs. "What's the beaver count on your hat, Bobby? I've always wondered."

Then Bobby starts talking politics.

DURING A LULL in the work on my first rig move, I had been chatting in the dirt with Tex, and he asked me who I voted for in the recent presidential election. It was less than a year since Obama beat Romney and gained a second term. Tex had asked in an offhanded manner, assuming I'd say Romney, and then he'd make his point. But I am not a conservative.

I had just moved to a very conservative state, and I wanted to make friends, not create problems for myself. I took a moment before answering. I had to really chew on it. I had made a promise to myself when I got to North Dakota, and it was very important to me: while I felt no compulsion to share much in the way of personal information, I had decided that I would not lie. This was the first time I was truly tempted to break that promise. It was my first real test.

"Obama," I told him.

"Jesus fucking Christ," he said, "you're a fucking Democrat!"

I didn't consider myself a Democrat at the time. I voted independent of party and I tried to explain this, but Tex didn't care. He was on his phone in a second; I was seized with sudden anxiety. "Please don't spread it around," I begged him. "I don't want to deal with this."

"I won't tell anybody!" he waved me off. "Hang on a second." Then into his phone he said, "Hey, Bobby, did you know Mike is a fucking Democrat!?"

About a hundred feet away, I watched Bobby Lee, in all his denim glory, turn to look at me with his phone up to his ear.

"Oooh, man," I said. I had only just started work, and I was struggling—not only to not completely fuck up but also to be, in any small way, acceptable to the people around me.

"Mike is a Democrat!" Tex repeated, and laughed a big belly laugh. He got off the phone, all but wiping tears from his eyes. "I don't think I've even ever met a Democrat!" he told me.

"I'm an independent," I said lamely.

"You're a Democrat," Tex said.

A roughneck was walking past us. "Hey, buddy!" Tex hollered. "Did you know we got a fucking Democrat on location?"

The roughneck stopped in his tracks. He turned and looked at me silently, sizing me up. He didn't smile. "We've already had two accidents on location," he said. "There might just be a third." He walked away.

"Jesus Christ," Tex said, "that guy sounded serious."

"He did sound serious," I agreed.

Bobby Lee had walked across location and quietly sat down beside me. "We're all here because we're capitalists," he had counseled me like a patient father. "Obama is a socialist. He wants to give your money to crack addicts."

"You know," Tex chimed in, "Obama's health czar wrote a book about population control and euthanasia. And they are going to use the health care law to euthanize people who are too old to pay taxes. You got a grandma, right? What do you think about that?"

"I think that's not true," I said.

"Jesus Christ," said Tex, "a real flesh-and-blood Democrat!"

Word spread fast. Tex took particular delight in outing me to swampers and operators while I stood silently next to him. At least he was funny about it. His excitement truly seemed to come from having never met a person with a different political opinion. "No wait," he had told me, "there was one neighbor we had growing up in Texas. He was a Democrat, I think."

"Two unicorn sightings in one lifetime," I said. "Amazing."

None of the younger guys cared much one way or the other. From what I could tell most of them figured they were supposed to be Republicans but hadn't spent much time thinking about it. "I don't even really know what I am," Huck once said to me.

"You should be a Democrat," I counseled him, slyly.

But the older guys did care. Bobby Lee was the worst, with the Wildebeest coming in a close second. This was ironic, since the Beest, like so many men I worked with, was a felon and therefore ineligible to vote in the state of North Dakota—he had very possibly never cast a ballot his entire life. One morning in the van, driving to location with Bobby Lee and a group of older drivers, the Wildebeest said, "You didn't really put your vote on that motherfucker, did you?"

I told him that I had, and an immediate uproar ensued. "A mile off location would be a good place to dump a body," one of the truckers cracked. It was a little nerve-racking. The midwestern sense of humor is a dry one, and it wasn't always clear to me who was kidding. I had to laugh it off, though. I put on a deep movie-narrator voice and said, "Michael was last seen heading south in a van full of Republican truckers." It got me some chuckles, and my resistance may have earned me some respect. The difference in political leaning seemed to, in a weird way, bring us together.

But Bobby Lee needled me about it relentlessly. It got under the skin of the calm, cool, collected truck pusher in a way that no other thing did, not that I saw. It was the only thing that penetrated his unflappable demeanor. He claimed to have been all but financially ruined by Obama's tax policy. I had no reason not to believe him, but I didn't want to talk about it, either. Not all the time. And definitely not while harnessing up to climb the derrick floor with a hammer in my hand. I was trying to keep my head down, get good at my job, and fit in. I was not trying to debate tax policy.

AT THE BAR in Glendive, Bobby starts harping on me again. "Why are you in the oil field if you're voting for a socialist?"

He encourages the gathered hands to gang up on me. "Yeah, how come you're in the oil field when you're a socialist? Huh? Yeah, how come?" I understand that the guys are playing along for sport, but Bobby Lee won't let up.

"I just can't fucking believe you voted for Obama," he says.

"Yeah, why'd you vote for him? I can't believe it either," the hands parrot.

"I voted for him twice," I say. "And I'd vote for him again, you dumb motherfuckers." My declaration is met with a chorus of groans and laughter. I'm the only Obama voter probably in the whole bar, almost in the entire state. And now I'm drunk and getting pissed off.

Bobby Lee is drunk, too, more than me. He's unsteady on his feet. "Keep that motherfucker away from me," he barks while moving closer to me. "What about the children, Mike? What about the children?" Bobby is a close talker. He cocks his head to the side in a way that makes you feel like he might go in for a kiss. He gets too close to me. I can smell his breath, and I don't like it. I think he might start crying. Or he might take a swing at me. I almost hope he tries it. This is my boss. But fuck it, I'm fed up. "What about the children, Mike?"

"Fuck the children, Bobby."

"See," Bobby says, turning to the other hands, "that's how a liberal thinks. That's what they think of this country."

"Fuck this country, Bobby."

"What?"

He's turned on me now. He swings around, and we square off. "What did you just say, you sonovabitch?" All pretense that he's joking is gone, and the guys behind him sense the change in the mood. They grow suddenly quiet, listening. "What did you say about America?"

"Fuck America," I tell him. A roar comes from behind. I've taken it too far. The guys will put up with a liberal socialist communist, sure, but they don't like this kind of talk. Bobby peels off from me and eyes them up. There's real annoyance coming from the men. Half a dozen of them stand in a tight half circle watching me.

"Fuck America," I say it again, defiantly. *I'll take all of 'em on,* I think. I'm drunk and I'm pissed. I'll get my head stomped in, but fuck it. I'm ready to go down swinging. Fuck this country they talk about, this homogenous, xenophobic, racist, backbreaking hellhole, where a man has to travel 2,000 miles to get a job. This isn't the country I was raised in. This isn't the country I love. This isn't the ideal I dream of. Fuck this America. The America I love is in Sunset Park, Brooklyn, where you can hear seventeen different languages as you stroll through the park. Where the music spilling on the street comes from Mexico, the Dominican Republic, Puerto Rico, Pakistan, Afghanistan, and Greece. Where on a clear night at the top of the park, under the big tree there, you can smoke joints and stare at the Statue of Liberty in the distance. That's the America I wake up for, that I go to work for. And I'll fight for her right now, you dumb sons of bitches. You can kick in my ribs, black my eye, and knock my mouth loose, and I don't give a fuck. I'm fed up with this. I'm sick of getting humiliated by a bunch of fucking hillbillies. Let's do this.

"Let's do some fucking pickle backs!" Porkchop breaks up the scene with a tray of whiskey and pickle juice shots. As quickly as the tension hits a fever pitch, it disappears, blows off like dust across location, dissipates like the morning fog over the kettles. We aren't politicians, we're oil field hands. We raise our glasses, clink them, and slam them down.

Bobby grumbles angrily, he holds his cowboy hat loose in one hand, a bottle of beer in the other, and wanders outside for a smoke. I lean back on a stool and catch my breath. I'm wrong, of course. My inarticulate hollering only confirmed these guys' suspicions—the ones who give a shit anyway—that liberals hate their country. I don't hate America. Not mine, not theirs, or anybody else's. I'm just angry and scared and lonely. "Fuck America" was the most unholy oath I could hurl. It was the ugliest thing I could say. A meaty, tattooed hand rests on my shoulder. Porkchop. I look up to see him staring down at me, a wide smile stretched across his broad face reveals uneven rows of pointy teeth. I have to laugh.

"I'll have another beer," I tell the waitress.

"GET UP, PUSS-PUSS." Seven a.m. the next morning, I blink my eyes open. The Wildebeest sits on the edge of his bed in a pair of tighty-whities. He looks nine hundred years old.

"Oh God," I say. I roll out of bed and rub my face. My head is pounding. I pull my pants up over my boxer shorts, walk into the bathroom, and splash cold water on my face. We are running late. I dress quickly, and we head out the door.

"You all right to drive?" the Beest asks me.

"I'm fine," I say.

"Are you sure?" he asks. He's in rough shape.

We get to location, somehow beating most everyone there. Slowly, trucks start pulling in. "Anybody, uh, know the way to Glendive?" Smash cracks over the company radio.

The hands start getting to work, pale and tired and wasted. Porkchop in particular is a mess. He has to walk behind company housing to throw up. He can't even keep water down. I give him a bottle of Gatorade and tell him to sip it carefully.

I feel rough, but I feel okay, too. There's a hangover licking at my heels, but I can outrun that. I've done that a thousand times with a guitar in my hand, or with a shovel, at a desk, or construction site. I watch the bedraggled hands nursing heads and bellies as they meander around location, bullshitting, rehashing the previous night's party, putting the pieces back together with sheepish grins, and I think, *I can outwork these motherfuckers.*

The sun begins to crest the eastern hills. New York is over there,

I realize, watching those first yellow fingers reach up over the dirt. *Fuck New York*, I think. The sky in Montana is a thousand miles long. It yawns above me, and I stretch beneath it, pulling my arms and legs loose, rubbing my hands each one with the other, opening and closing my palms from fists to jazz hands and back again, feeling my blood start to move through them.

This is my day. I can feel it in the movement of my blood. I'm going to bang this day into shape with a hammer, bend it to my needs with chains and shit hooks, fashion it to the construct of my own will. I'm gonna put my hands on this day.

"Why'd ya go to South Dakota anyway, Michael?" I'll hear the question the moment I decide to leave New York, and for years after, from what feels like a million different mouths.

"North Dakota," I'll say.

We all agree that life is suffering, right? It may be the one single idea that unifies Buddhists, Christians, Muslims, Jews, and atheists. Life hurts like a bitch. It is *supposed* to hurt like a bitch. It was built that way. Does that make the world a people-breaking machine? I don't think so.

"Why did you go to North Dakota, Michael?"

Because if life is suffering, it must be worth suffering. Because if life is to have any meaning, I need to take responsibility for it. And if life is suffering—and it is—then I need to take responsibility for that suffering, too. Don't I?

Why? doesn't take you out to lunch for a cheeseburger with all the fixins just how you like it. It doesn't salt the rim on your margarita or rub lotion on your hands and ask you how your day's been. Like life, *Why?* hurts.

The scars in the land on which I toil. The scars on the body in which I work. The scars across my very soul. The father wound in my heart. My grief for my sister running like a gully through the whole length of my very life. I must be responsible for all of it. I can't go on. I'll go on.

I came here to do the hardest thing I could find to find out if I could do it. And I do. I get to work. And I do. *I outwork all these motherfuckers.*

HUCK AND BLOOD

Went to Sidney to see some country guys at the fair, Stan, Gibby, and me," Huck tells me, naming two of his best pals. "One singer was that 'Tequila Makes Her Clothes Fall Off' guy. You know that song?" Here Huck sort of sings a line: *"Margaritas at the Holiday Inn . . ."*

"Yeah, I know that song," I tell him. It's a total earworm. Along with Blake Shelton's "Boys 'Round Here," it is a ubiquitous part of Williston's summer playlist, blaring out of jukeboxes in every honkytonk and through the satellite radio of just about every other Dodge Ram pickup that cruises down Main Street. Two other songs on repeat that summer are Katy Perry's "Roar" and Miley Cyrus's "Wrecking Ball." Watching Miley swing naked on a crane with a group of actual crane operators is a very particular experience; and rarely have I laughed as hard as I did as when riding in a van with four oil field hands belting out *You held me down but I got up* as Katy Perry took a shower with an elephant on one of their iPhones.

"We were watching the music," Huck tells me, talking about the country music at the fair. "And Stan and Gibby and me end up walking out to the truck for some reason. We're drunk. Gibby is taking a piss and he ends up pissing on Stan. Like hitting his leg or something. Stan threw him on the ground, and I grabbed Stan. Now, he's a big boy. Stan's been in construction his whole life. He starts freaking out, and I realize he's gonna kick my ass if I let him go, so I throw him on the ground and hold him down. And he puts his finger in my mouth on this side, and he fucking ripped my lip off my gums down here! He pulled it off like that far." Huck pulls his lip out from his gum to show me. It separates to a weird degree.

"Ouch," I say.

"So I get up and I'm like, 'Oh my fucking God, dude! Jesus!' I was tasting blood. I was like, 'You sonovabitch.' I hit him and I hit him and I hit him. I hit him like six times. But the first one was right in the jaw and he fell down, hit his head on the grille of the truck. And then I hit him in the side of the head, and that's when I started breaking my hand. My fucking hand was swoll up for like two weeks, man. Back here was like"—Huck shakes his head and emits a nasty sound—"and I couldn't really use this pinkie for a while."

"Fun stuff," I say, not smiling. Huck looks wounded for a moment at my response. I'd usually make fun of him for his fighting. I once told him and a group of swampers, "I spent my twenties making love to beautiful women." I stretched out the words to be as obnoxious as possible. "You dumb crackers are spending your twenties kicking each other in the head."

But I could see how these battles between Huck and his buddies brought them closer together. "Stan's got daddy issues," Huck would say. Throwing each other on the ground was a way for them to touch each other, to hold each other. It was a rough love that mimicked the love they got from their fathers, but it was also wholly their own. It bonded them. In some ways, it was just roughhousing—

but after a roll in the dirt they'd have a story to tell that solidified their bonds, made them more of a crew. It also drew a line between the guys they'd busted up with and guys they hadn't. In the form of black eyes and swollen hands, they marked themselves with this love, visibly showed it off to the world. I could judge Huck and his buddies as much I wanted to, but I'd see it again and again. Fighting made them tighter.

And my disdain for it was hypocritical. Me and my best friend didn't go out punching dudes when I was in my twenties; instead, we chased women. We were as relentless and myopic about it as Huck and his buddies were about their fistfights.

Huck reminded me a lot of one of my lifelong best friends, Seamus. We'd known each other since we were teenagers. Like Huck, Seamus was a fat kid growing up, and he seemed to carry that slight insecurity into adulthood. Seamus studied acting with me, but like me, he found it difficult to think of himself as an artist. After college, he rotated between working as a motorcycle mechanic and picking up acting jobs. After the two of us went through breakups, we spent all our time chasing girls.

We were boys in our early twenties, and our nights out were as much about us spending time with each other as they were about engaging with the opposite sex. When we did meet women, we worked to charm or rattle them. We pushed their boundaries, teased them. Some ignored us but some played along, and some teased us back. It was a lot like the buildup to a fight in that way. There was a sizing up, and a sort of circling. When Huck fought, he'd take his shirt off. Our goal was always to take our clothes off.

Seamus and I slept with as many women as we could. We slept with the same women, sometimes at the same time. Like Huck and his pals rolling around in the dirt, it drew us closer together. The desire must have been connection—not just with them, but also with each other. With women, the reality of the connection we pursued was structured as much around power as intimacy—like the strangers Huck and his buddies fought. We weren't innocently trying to get closer to them; we were asking them to submit to us.

And of course, many of those we pursued were not charmed by Seamus or me at all. But some of them got a kick out of us, some of them I'm sure got the same power rush we felt from sex, and some of them ended up becoming real friends. Others, I know I hurt. Now, I find my actions irresponsible and my motivations sad. I'm not proud of this behavior, but I sure was proud of it then.

For post-adolescent boys who haven't been taught any better, the desire for, as well as the pursuit and result of, both fistfights and sex can be surprisingly similar.

"Everything was fine after that," Huck assures me. "After Stan got up, he was like, 'Fuck man, I'm sorry.' I'm like, 'I'm sorry,' and Gibby's like, 'I'm sorry.' We looked like we'd gotten in a fight. There was blood on us. We just walked back into the fucking fair, though."

Blood. It always came back to blood. It was in some way as if Huck's battles with his buddies were an attempt to bleed out the father wound, to empty it or to clean it.

Huck had "Ámharach" tattooed on his left forearm. The word is Gaelic for "lucky." A jagged scar ran through the letters as if the word had been crossed out.

"How'd that happen?"

"We were at Biff's, me and Smash and a few others, and I punched through Smash's topper window on his truck. Smash punched through the topper window on my truck. I got this scar and Smash got cut by the crook of his elbow. Biff ended up punching me in the face that night. We had to go to the hospital and shit. It was right after Biff's wreck."

"After the Fourth?" I ask.

"Yeah. He was on crutches," Huck says. Biff had wrapped his car around a tree on Independence Day. "We were drinking. Krystal was there. She was freaking out about something. Logan wasn't there, which is why she was with me."

Huck's honesty could be bracing. He knew Krystal was using him, and he never tried to sugarcoat it or paint himself as the winner.

Huck continues, "It was like shots, shots, shots! Crown Royal, Crown Royal! Pendleton! Fucking Wild Turkey! It was all kinds of shit . . . beer, too. I got really fucked up. Me and Biff got into it. I don't know what started it. Biff just fucking popped me out of no-where. Well, not out of nowhere. We were like, standing shirtless looking at each other like, 'Fuck you! No, fuck you!' kinda thing."

"That'll do it," I say.

"And he just fucking hits me." Huck says, "I slammed him to the ground and then I crawled on top of him." Huck mimes realizing Biff's leg is broken. He's surprised! "I was like, 'I'm not gonna kick your fucking ass, because you're fucking hurt. I'm not gonna do it.' And then I let him up."

"And he was like, 'Well, we can go do this again in the grass.'"

"And I was like, 'All right, whatever, let's go. I'm gonna fuck you up.' It never happened. We went out to the grass and had a warm embrace. 'I LOVE YOU, BROTHER!'"

Huck and I share a good laugh, me shaking my head in mock woe and Huck delighting in his own bold acts of stupidity.

"The scar through your tattoo?" I ask again.

"After that is when I punched through Smash's topper," Huck

says, although he can't quite remember. "Or before. It had to be after. Or before. Because I had to be all bloody before. We all had our shirts off. Smash and I were putting our blood all on each other. Blood Brothers! I should've got stitches."

"Ugh," I say.

"Smash got cut up by the crook of his elbow. Every time he moved his arm, blood would spurt out." I make a face, and Huck continues, "Oh yeah, it was bad. And then Biff is like, 'You been bleeding for a while, man.'"

"Smash was like, 'Yeah, dude.' And he got light-headed right after he started thinking about it. So I threw him in the truck." Huck gets taken away by his own telling of the story, miming taking the wheel and shifting from first to second. "I was driving, no license, drunk, driving like Dale Earnhardt or something. I was going sideways around corners. Smash is going in and out of consciousness. I'm fucking hauling ass. He noticed enough to wake up and be like, 'Fuck, dude, slow down.'"

"And then we get into the ER, and it's the same doctor from jail, who pulled back the foreskin on my dick when I was processed. I was like, 'Hi.'"

I shake my head while Huck sticks a cigarette in his mouth and lights it. He squints, ruminates for a moment before adding, "That was 'cause of Krystal, now that I think about it. She started something and then Biff and I got into it. I forgot about that."

A NIGHT ON THE TOWN

"W hy do men like women with big titties and small pussies?" she asks me. She's in her midfifties, with short dark hair, lots of eye makeup, and a pair of ill-fitting dentures. She's drunk, slurring her words, nearly falling off the bar stool, while the teeth nearly fall out of her head. "Why do most men like women with big titties and small pussies?" she asks me again.

"Because most men have big mouths and small dicks," I say. I have no idea where I'd heard that before; maybe I just knew.

"Heeeeey!" she says in affirmation. She holds her hand in the air until I give her a high five. Then she turns to the two Minnesota dudes I'd been talking to, and she high-fives them, almost missing both times.

"You gotta look at the other person's elbow," I offer, but no one seems to hear. I'm at DK's Lounge. It is evening, and I have the next day off work. Party time.

The place is mostly dead, but I'd been talking to the dudes from Minnesota for a while. They were in their midtwenties, dressed in shorts and polo shirts. They had been golfing that day with friends when they met two young women in the clubhouse. Thinking they had a shot with them, when their buddies decided to leave, they stayed. But the women took off soon after and the dudes were left without a ride. So they walked from the golf course to the bar. I was sitting by myself when they came in. They invited me to join them and, after a shot of tequila, told me their story. It was a story that belonged to a normal town: Friendly guys flirt with friendly gals but end up hoofing it home alone. It was so out of place in Williston. So were the dudes. They seemed untouched by the boom with no pretension of toughness. They could have been on vacation in Florida.

When the woman showed up, they took an interest in her, asking questions, telling jokes, high-fiving. Too much high-fiving. I wanted her to leave. She was a drunkard, ugly wasted. She looked like she'd been run over by a car, messy and slurred. Damaged like the town. Desperate like the town. She belonged here. Even her voice was booming.

As we sit there, another fella, maybe twenty-three or twenty-four years old, slides up behind the lady, and slips his arm around her. He's thin and tall with styled hair, and he sips his beer like he's practiced it in the mirror. Maybe he's had a drink or two, but he is far from drunk. He ignores us men, smiles at the lady, looks deep into her googly eyes, and whispers something in her ear. I see his lips nuzzle her ear. She laughs a loud barking laugh, and he moves in for a kiss. In an instant, the two are engaged in a sloppy make-out session right next to me. I scoot away, but my eyes get stuck on them. I nearly toss the contents of my stomach. The Minnesota dudes order another round of tequila and head quickly out the door.

It bums me out. I finish my beer and use the bathroom. When I walk out of the bar, the lady with the fake teeth and the young guy are in the parking lot, cigarettes burning idly in their hands, wrapped in each other's clingy embrace, making passionate suck face.

I drive downtown with my own lonesomeness. I'd been trying over the past weeks to lure Huck out for a beer, but with no luck. I was alone all the time. It was tough. I had a small group of people I kept in touch with through emails I sent on days off, music pals, women I'd dated or wanted to date, a couple of actors, my brother Matthew. I spent hours writing down what was happening to me, all in an effort to connect. My emails kept growing longer. I could feel all this loneliness wearing on me, making me weird.

I park outside the boardinghouse and stroll farther downtown. It's past 10:00 p.m., and I can't think of the last time I've been up this late. I walk down East First. The night is warm, and a breeze filters through the streets—the pungent stink of blooming flowers, the ripe smell of freshly cut grass and greening leaves. I get a sense of what Williston must have been like before the boom—the sweetness of the air, an occasional whiff of petroleum. It is peaceful, uncluttered and serene. For a moment, it even feels safe. At the corner of East Broadway a Mack truck rumbles by.

In a motorcycle bar called The Shop, the bartender asks to see my license. "I wouldn't usually card ya," she says, "but the cops are looking for somebody, so I gotta check all the IDs in case they come in and start asking around."

From there I head to a bar called the K K Korner Lounge. The name of the place terrified me when I first saw it, but like so many things in Williston, I'd grown used to it, accepting the lame explanation that it was nothing more than a reference to the names of the two owners. Like DK's Lounge and The Shop, the bar is mostly empty. I had figured, before moving to Williston, that even on off nights, downtown would be white hot with wild young bucks burning through the cash in their wallets. But it is a weekday, and the city is hollow.

There is one group in the place. Past the bar at which I sit, a tall, slender young woman with long brown hair dances with a small group of friends. She pauses to talk to a couple bored-looking guys sitting on couches and sipping beers with their caps pulled low, and over the din of music, I detect an Eastern European accent. I look at her, and I'm immediately filled with lust—those long legs, that slender waist. She is beautiful by any standard. It's been months since I've been with a woman, since I've touched one, since I've even talked to a woman I felt any attraction toward. *She must be a hooker*, I think. I look away, and it strikes me how messed up I'm becoming, and in such a short period of time. I have no idea who this woman is, what she does, or where she's from. I've been in a boomtown for ninety days, and I'm ready to assume every woman I see is a prostitute. What the fuck is happening to me?

HUCK AND MR. HYDE

As summer had faded, Huck's goofball streak took a serious turn south. Deep in the throes of his passion for Krystal, he continued to throw back drinks with her boyfriend, Logan. Like two gorillas competing for the attention of a mate, they kept making fools of themselves and getting in trouble whether she was around or not. Animosity between the two would start as competitive drinking and attempts to out-macho each other, but it began to seriously escalate.

This was the beginning of what Huck would later call his Mr. Hyde phase. He'd drink to excess. He'd black out. Then he'd become a different person.

At DK's Lounge, Logan gave some guy he'd never met $400 to buy cocaine. The guy left, and Logan and Huck got drunk waiting for him to return. The guy never came back, of course, but it was hours before Logan could accept that he'd been ripped off. When Logan and Huck finally left the bar, they were blind wasted and talking about bashing the guy's head in.

"Logan had just gotten this new 6.0 pickup, and that has a lot of power for a gasoline motor. They're torquey and they fucking break 'em loose," Huck tells me. "We come hauling ass outta DK's, and Logan whiskey-throttles it right into a retaining wall. Blows the whole front wheel off. And he doesn't know it. So by the time he gets turned around, he's driving on the fucking brake.

"I tell him to stop, and we pull over by the lumberyard. So we get out and we start walking to the back of the truck, and he just starts punching me in the chest, hard. Like as hard as he could, boom-boom-boom. I'd kicked his ass before. I'd kicked his ass like a month before, so I let him hit me. I say, 'You feel better now?'

"He was like, 'Fuck no!'

"He called Krystal. She drove over to us, and we rolled the truck tire from where it had fallen off all the way back to his truck. Meanwhile there's the CV axle, brake assembly, ball joints, everything's still in that wheel. The thing weighs like three hundred pounds. We both grab one side and we go to put it in there, and he bows out halfway up, just lets go. So, I'm holding on to the tire, and it comes shitwhipping down on my knee. Three hundred pounds of fuck-you sliding down my leg. I just fucking snapped.

"I don't remember any of this—Krystal told me later—but I picked Logan up in the air and threw him on his ass like 'Roooaaarrr.' He got back on his feet and ran to Krystal's car. I walk after him. I'm getting ready to lay hands on him, and Krystal's like, 'Huck, chill out!' But I'm just like the fucking Terminator, dude. Mr. Hyde.

"They take off. And some guy comes out of the hotel there and he says something to me and I just threw him to the ground. I was like, 'Fuck you, dude.' Then I grabbed my guitar outta the truck, and that's when the cops showed up. Right when I was trying to walk home with my guitar.

"And they said, 'You can't leave. We gotta ask you some questions. We had some complaints.'

"I kept telling them the whole time, 'I'm not trying to be rude to you guys. I just went through an extreme situation.' But then the guy who I threw on the ground said something, and I was like 'Fuck you, motherfucker!'

"And they were like, 'All right, that's disorderly conduct. You're going to jail.'

"That's when they grabbed me. They went to push me up against the cop car, and I started struggling with them.

"They were like, 'I'm gonna tase you! I'm gonna tase you!' And I was like, 'Fucking tase me!' like I'm going crazy. Krystal saw it all. She'd driven back around. I kicked some of them in the balls, I guess. I should have got assaulting a police officer. And a lot more. And they should have kicked the shit out of me.

"They put me in the car and they closed the door, but they couldn't get the hobble restraints on me. I'm way too tall for them. So I started kicking the door as hard as I could. I kicked it so hard the door started bending out and separating from the frame. I was Hulking out bad.

"And then the cop was like, 'Quit kicking my door!' And that's when I spit in his face."

"Jesus. Do you remember that?"

"I'd come out of it for a second, and then I'd just go back into it. I don't really remember much. I remember being in the back of the car and feeling this rage. Felt like I was in a dream, kinda. Alcohol dream. Zombie. I think that's what's gonna get me out of it. The fact I'm an alcoholic."

MAGIC

G et over here, Magic Mike!" the Gruff Crane Operator yells at me. "Help me tie this down."

I grab a boomer and a chain, and we work side by side for a moment in silence.

"You know that's what they call you now?" He grins sideways at me, even his Oakley's seem to smile. "Magic Mike?"

"Oh yeah?"

"Oooh, yeah."

"Huh."

I pull the chain tighter and crank the shortener.

"Over the radio, you know," he says. "A lotta chatter."

"I've never seen the movie," I tell him. A popular film about a group of male strippers, it had come out the year previous. "But I do have some pretty good dance moves. Funny you guys just kinda knew that."

"I don't think so."

"Like intuitively."

"Not at all."

"Oh yeah, I can really get it done on the dance floor."

"That is more than I ever need to know about you," he says.

"I could show them to you sometime . . ."

"Unnecessary."

"I mean, not here on location. That would be weird." I smile at him, "But we could go someplace more private."

"I don't want to do that."

"I've got this really great pair of pants . . ."

THE VIKING

It is foggy as we ride north to location. At times, I can't see more than a few feet in front of the truck. In areas where the road clears, I look out across the landscape. Every small pond and lake looks like a cloud, thick as cotton, has settled on top of it. Like the kettles are wearing magic hats.

At the job site, Huck takes the green hard hat off my head and throws it hard into company housing. "Fuck this fucking greenhorn shit," he says. I run after it. When I return, Huck gives me a spare white hard hat belonging to his operator. "Wear this," he says. I feel guilty putting it on. I've only been in the field for about three months. I am getting better at the job, but I don't want to overstep my bounds. Getting my ass handed to me for wearing a white brain bucket would be embarrassing.

I hop in a haul truck with Olof Fjeldstad—a voluble truck driver who talks in a mixture of trucker slang, curse words, and mechanical arcana in a low, winding growl that I never truly comprehend. At thirty-seven years old, a strapping, six-foot-four tall Adonis with blond shoulder-length hair, chiseled features, and a pronounced chin, Olof had good looks and a sturdy frame that had earned him the nickname the Viking, but he isn't intimidating. He has an odd-ball sense of humor and an infectious grin that stretches across his face like the Cheshire cat's. I had met him on my second day of work, and I immediately liked him when he launched into a story about feeding his dogs the sandwiches his wife made him for lunch. "They're good dogs, goddammit. And she's a good woman."

I got the sense that he liked me, too—I was another oddball—and that was a rare, precious feeling, especially in those early, friendless days.

"Damn, Magic Mike! You're blinding me with that new hat!" the Viking says.

"Huck promoted me," I tell him.

"Huck did? Well, take what you can get."

Fjeldstad broke his collarbone almost as soon as I started at Diamondback. At his family's Fourth of July party, he wrecked a 4-wheeler while hot-dogging through an oil rig on his father's farm. "Thing went sideways on me and I woke up on the ground, god-dammit," he said.

After the accident, he went back to the party and opened a beer. "Don't tell anybody," he growled to his buddy who'd witnessed the wreck.

"When he saw me flip that goddamn thing, he thought I was gonna get up . . . never," he growled to me.

It only took a few sips before he realized something was very wrong. He was going into shock. His mother-in-law took him to the hospital. "Didn't shed a single tear, though," Olof said.

I was disappointed when I heard about his wreck. Now, I'm just happy that he's back. And back he is, talking trucks like a stream over a broken dam, me sitting swamp side playing chase the rabbit with his jargon, catching just about every third word: *carburetor, transmis-*

sion, goddammit, winch pole, hydraulics, goddammit, gasoline, brake fluid, John Deere, goddammit. I'm riding on the rhythm and spirit of his monologue, whatever the hell he's talking about, anyway.

The haul truck on this move is being used like a bed truck, transporting storage units, engines, and generators from location to location and dropping them in the vicinity of the crane. The Viking wears a sling on his arm so he can drive, but he can't throw his own bridles, so I'm put in the haul truck to do that for him. It's as easy a job as I've had since I started. We have a fine time bullshitting as we go.

Late in the day, Fjeldstad backs the haul truck—trailer and all— up a hill and through three ninety-degree turns within what is maybe a two-hundred-foot stretch of dirt road at full speed. I watch the Viking with one hand on the wheel, one arm in a sling, his jaw set, his head swung back. We fly through the turns, the trailer twisting at angles just barely possible—the geometry of a trick pool shot done at high speed with an 80,000-pound piece of diesel-fueled Peterbilt. Fjeldstad slides the truck easily in place, lined up perfectly next to another driver's trailer.

"Well, goddamn!" the other driver hollers. "The fucking Viking!"

AT THE END of the day, before we leave location, Huck's operator demands his white hard hat back. I give it to him and retrieve my greenhorn. Sheepishly, I carry it back with me to Fjeldstad's haul truck.

We ride back toward the yard, driving through an endless maze of construction on the way north past Watford City. "There are two seasons in North Dakota," the Viking says. "Winter . . . and Road Construction, dammit."

Driving, the Viking notices a single orange safety cone lying on its side. "Got one laying over there," he says, and points it out to me. As we pass, I watch in the mirror as he kicks the back end of the trailer out so that it just . . . barely . . . nudges . . . the orange cone back to standing. We cheer.

"I'm glad you were here to witness that," he says, surprised by himself. "Nobody'd believe that!"

There's a pause as we continue riding north. "You're all right," Olof says to me. "You're all right with me. You're a good hand."

It is a stretch for him to say that. The words sound funny in his mouth. But maybe it isn't that far from true. On the Glendive move, Tex had said to me, "When you started, I didn't think you were gonna make it. Some guys you can just tell. But I saw you out there today, and you were making a hand."

"I'm getting there," I tell the Viking, and he nods.

When we land back in the yard, we head inside and hang around the office, bullshitting with the dispatchers. Fjeldstad asks them about giving me a white hard hat.

"How long you been with us now?" a dispatcher asks me.

"Just over three months."

"Well, you're gonna have to wait six months before we can give you a white hat," he says. I shrug, and Fjeldstad steals a glance at me.

Back at the boardinghouse, pulling off my work clothes, I get a text from the Viking. Well, goddammit, look at that. He swiped me a white horn.

THE FIRST PICK

No matter how swell or simple or lazy one workday may be, it is inevitably followed by the cold, creeping feeling of dread that accompanies day one of a new rig move.

I hop in the truck with Bob Olhouser and the Beest. The Wildebeest is telling Olhouser about the move in Glendive. As we pull out of the yard, he invites me into the conversation with a question about the crane, and I mumble an answer. It is early morning, and I'm still pulling the cobwebs out of my brain. Later in the drive, the Wildebeest turns around to joke with me about Porkchop's hangover. We share a laugh, and I realize this is the first time the old man has included me in a conversation when anyone else was around.

The Beest and I are the only two on the new end. The locations are out of sight of each other but still close together. We'll have idle periods, but when the trucks show up, which will be frequently, we'll need to unload them as quickly as possible. There are five haul trucks on this move. At any moment during the day, four of those trucks could be waiting on us. The job is intended to go fast.

The day is cooler than any day I have worked. All summer long the needle seemed stuck in the 90s. But it is the second week of September, and the temperature has dropped like a hammer: 74 degrees and breezy, an incredible reprieve. For the first time since I moved to North Dakota, I watch the sun climb over the plains looking like a pal.

A truck pulls in, and I quickly unload it. The lease is smaller than most, and I have to direct the truck around a tight corner to get it pointed back out toward the road. The next haul truck arrives almost immediately, and I jog across the dirt patch to meet it there. We unload it in silence—unhooking the boomers and chains from the trailer, then attaching the gin truck's hooks to the metal and lifting it off, walking the load to the edge of location, dropping it and repeating. Alone in the dirt and at peace with the work, I watch the wind play with the prairie grass surrounding us, a rumbling, tumbling, splashing ocean of grain.

I jog back to the haul truck as another one enters the access road to the lease. I realize that I'm responsible for rigging down everything that some six other guys rigged up on the old end. I don't know why the move was set up that way, but I don't mind it. I like the feeling. I'm outworking all of them. My body moves against the cotton of my jumpsuit, the scoria crunches under my feet, the breeze kisses my sunburned hide and cools it. I climb up onto the waiting truck's trailer and start throwing chains.

The work, for the first time, is in my hands. It is in my blood and my bones, the sinew and muscle of my chest and arms. I am an extension of the work itself, a tool, a vessel born to complete a task, and I feel simple like goodness.

. . .

BY AFTERNOON, I am all but dogged out. I overexerted myself, got carried away with the breeze. But it is a feeling I'm comfortable with now. I'm used to being tired. So, I keep working. My hands and legs keep moving, keep giving to me. When it comes time for a break, I hop in the cab of the gin truck. The Wildebeest has the music cranked up. "Call Me Maybe" is blaring from the speakers.

Hey, I just met you and this is crazy

But here's my number, so call me, maybe?

It takes me a moment to establish my bearings. I look at the fat, dip-chewing trucker next to me with his white mustache, cap pulled down, and space-blue sunglasses wrapped around his head, and I say, "What the FUCK are you listening to?"

IT HAS BEEN a good day, a long day, and my limbs are loose and weary sitting in the cab, the sun slanting through the window. A haul truck bounces up the access road and enters the lease: the last load of the day. I'm about to hop out when the Wildebeest stops me.

"You wanna operate the gin truck?" he asks.

"Do I?" I say, "Hell, yeah!"

"All right then." He climbs out onto the scoria and hollers at the truck driver, "You mind throwing the bridle? I'm gonna show him how to operate the truck." The driver doesn't mind at all. He hits the air brake, hops out, and walks into position behind the gin truck. The Beest climbs back into the cab and patiently directs me.

I gear the truck, working the transmission so the clutch is all the way in, the pedal pressed against the floorboards. I shift carefully into reverse and slowly let the clutch out. The engine beneath me wakens like a slumbering giant. I have my arm slung back and I'm staring out the back window, my eyes on the haul truck behind us. The gin truck lumbers in reverse, a wicked, metallic animal with grease for blood and axles for bones. The power under the hood is palpable in a way I have never experienced as a passenger. I haven't ridden horses since I was a boy, but the truck soon feels like an extension of me in the way that only a steed can. Only bigger. More powerful. More dangerous. I'm steering a metal war elephant. It's awesome.

As I look out the back of the winch truck, the perspective is strange. The haul truck driver appears incredibly small. It is hard to judge the distance properly. I stop the truck when the Wildebeest commands, then drop the chains. The driver throws the slings around the mud tank's feet, and, at the prompting of the Beest, I adjust the lever that pulls up on the chains. It is an incredibly delicate gesture that controls a machine of incredible force and power. I barely nudge the controls, and the tank is pulled forward, up toward the back of the gin truck.

The Wildebeest has me put the truck back in gear and drive it forward a few yards.

"Well, you just picked your first load," he tells me, his blue eyes shining.

THE HARVEST

A few days later, we're sitting in the truck and the Wildebeest says, "You need to get a new hard hat."

"Yeah," I tell him, turning my brain bucket over in my hands. "This one is pretty busted up."

"It's green," he says.

"Yeah," I say. "The Viking said he swiped a white one for me, but I haven't gotten it from him yet."

The Beest snorts. "If I were you," he says, "I'd grab that motherfucker."

Outside the truck, dust spills across the location like a tender tan tide washing eternally over a red-dirt beach. I look down at my feet, and it looks like the ground is moving beneath me.

I walk over to his truck and ask the Viking about the white hard hat. He's sitting in the driver's seat, and I'm standing on the ground below him. "You work Sunday?" he asks. "I gotta dig up potatoes in the garden."

"There's no move on Sunday far as I know," I say.

"Goddamn farming," he says, shaking his head. "Slaughtered a chicken last week. Made my daughter cry. She don't like eating the ones she knows, you know. But man, they are tasty."

"I'll come out, give you a hand," I tell him.

"All right," he says. He reaches behind him and pulls out a brand-new, unscathed, unused, bright white hard hat. He hands it to me.

"I'm looking out for you, Magic," the Viking says. "Can't have you running around like a goddamn greenhorn."

I leave location with the Wildebeest, feeling fine. We are in the winch truck, taking it back home over bumpy roads. He takes a dip, and I ask for a pinch. He gives me his can of snuff, and I drop a bit between my gum and cheek. It is tangy, sweet, and spicy, and it gives me a slightly nauseous light-headed feeling that I associate with a kid trying his first cigarette. On this day, on this bumpy road, under this sky, sitting swamp side in the Peterbilt, I like the feeling.

"Goddamn," says the Wildebeest, cracking an admiring grin. "You're gonna turn into a cussing, fighting, dip-chewing mother-fucker."

"I always thought I was pretty good at cussing till I came out here," I tell him.

"That's all there is out here," he says. Then he grows serious. "Sometimes I go to town, you know. I catch myself cussing in town. I don't like that."

I DRIVE OUT to the Viking's place, a house trailer set down on several acres adjacent to his father's sprawling farm. The Viking hands me a cup of black coffee when I arrive. We walk through the yard and into the garden, and we start digging, first using a shovel to overturn the ground, and then with our hands.

I reach down into the earth and pull out dozens of tubers, collecting them in a pile in front of each hole. The dry soil feels good between my fingers, the cold air tastes clean, and the sun shines fine on the back of my neck. I move down the rows on my knees methodically clearing each hole of potatoes, pausing to sip the strong, bitter coffee.

It feels good to work on my day off, to continue the toil, to not allow this connection I've made with my own body and this new dirt to grow slack. My hands have finally adapted to throwing chains. I massage them out of habit, but they no longer feel swollen or bent. I push them into the ground, my muscles moving beneath my shirt. My forearms and biceps have thickened over the past few months, my chest has expanded, and my shoulders have broadened—my body finally catching up to my stubbornness. Kneeling in the dirt with my hands in the soil, I watch my shadow stretch across the garden. I feel big, powerful, and resolved. I'm finding my place here. A cool wind rides under a hot sun. As the season is changing, so am I.

I thought I came to North Dakota to make money. I didn't. I came to become a good hand.

That evening, I call my brother Matthew. The last time I spoke with him, I was counting down the days before I could skip town. This time I tell him that I operated the winch truck. I tell him about the Viking's incredible job backing up the tractor trailer. I tell him about the white hard hat and digging potatoes. He pauses before he responds. He can hear the change.

"You're going to stay," he says.

"I am."

BOOK THREE

THE GOOD HAND

—

You've found something
Something I missed
You found a gladness being here
And how to stand up proud to laugh with everybody else
You found your work
And your notch
And where you belong
In a chain of others that can't be broken and a stone
foundation that can't be shook down.
So far, I haven't found that
I found a drifting wind and a blowing rain
And a coward and a stranger to people's pain
And people never will show you their laugh
Till you find it out through their pain
Maybe I'm learning
That secret of all secrets, (and it ain't even no secret).

—WOODY GUTHRIE

TWELFTH NIGHT

Wet, briny, plump, and sweet, an oyster slides down my throat; I savor it for a moment before washing the taste away with a swill from my rocks glass—hints of chartreuse and lime juice, a cucumber garnish, the sharpness of gin. Bebop saxophone roils the speakers above the bar, the sound of glasses clinking, silverware scraping plates and teeth, the chatter of animated conversation. Black-clad waitresses in tight dresses slip drink orders to mustachioed bartenders and walk off, hips swinging between the tightly packed tables. I slide my hand around the thigh of the woman I'm with. She meets my eyes with a sly smile, and I lean in for a kiss. My nostrils fill with the faint aroma of her perfume and the smell of her flesh.

"It's good to be back in town," I tell her.

ALTHOUGH INHABITANTS OF both places would be generally loath to admit it, there are startling parallels between New York City and Williston. Both are notoriously expensive. In both cities, real estate is an absolute obsession, renting a blood sport. Both are flooded by out-of-towners.

Both towns are job, status, and money obsessed. What do you do? What's your work? How much do you make? Where's your office? Where do you live? How many rooms? Do you have roommates? In no other city or town have I ever experienced these as the first questions asked by complete strangers upon first meeting. In New York City and Williston, that is the custom.

Both towns are oversized places not only in the imaginations of those who hear about them but also in the minds of the people who live there. Williston is the New Gold Rush and New York is the Broadway Lights. New Yorkers see themselves in the movie about New York. North Dakotans see themselves in the country song about the oil boom.

These towns are two arteries pumping blood to the heart of the American Dream. Go to the place. Get the job. Work hard. Rise up. Nowhere in America in 2013 is that dream more pungent than in New York City or Williston, North Dakota.

I HAD KEPT my apartment in Brooklyn when I moved to Williston, subletting my room for the months I planned to be gone. My goal was to move back to the city around this time, midautumn, to resume my life here. But things have changed. I don't want the city life anymore. I'm staying in Williston. So, I return to New York and sign over my lease. I move my things into storage. Then I play a couple guitar gigs to make some cash and have some fun. I gather with friends, and we go to the theater. Before returning to New York, I had $4,000 in the bank and $3,000 available in credit. I had no idea at the time—if you told me, I would've slapped you—but this would prove to be the financial peak of my oil field year.

WE ARE WATCHING *Twelfth Night* at the Belasco Theatre on Broadway. It is a Shakespeare's Globe production with the roles, like in the original Elizabethan staging, all played by men. The instruments used in the play and the music performed are from the Renaissance. Even the costumes are of that long-gone era, handmade fabric available in sixteenth-century England. In the middle of the grandest metropolis of the modern age, New York City, what I am experiencing is absolutely pre-petroleum.

In the first scene of the third act, Viola, dressed as a man, questions Feste, a clown, who is playing a drum called a tabor. She asks if, as a musician, he lives by the drum. He responds that he lives by the church.

Viola: Art thou a churchman?

Feste: No such matter, sir. I do live by the church, for I do live at my house, and my house doth stand by the church.

It is a four-hundred-year-old dumb joke.

My friends and I leave the theater and land back onto the streets. I'm with a married duo of documentary filmmakers, a red-bearded guitar player from Georgia and his girlfriend, a yoga instructor, my best friend in New York—a commercial voice-over actor—and his wife, a life coach, and my date, a creative director at a fashion company with a side hustle in rock and roll. We make up a real New York gaggle. We're on our way to hook up with a journalist I'd met when he interviewed me about a play I'd written and his best friend, a surfer–banjo player, who, last time I saw him, was living in his van.

Cabs burning fuel stream past. Electric streetlamps flicker through the coil of synthetic rubber wiring. The giant screens of Times Square advertise the world's riches: cotton polyester blends, food wrapped in plastic, music delivered through plastic earbuds. We escape the lights down into the subway, board and ride through the petroleum tube, back to the street and into a crowded bar. I notice the tabletop varnish made from petroleum derivatives, our glasses forged in oil, the fake leather seats petrol-made. We scan the laminated menus and order seafood caught on diesel-powered boats, stored in freezers, shipped in trucks, and brought to our table by waitresses in slinky clothes—only oil could make them coil so tightly around those bodies. And we swallow it, suck it, chew and devour it all—delicious, man-extracted, life-granting, world-destroying black gold.

I'm standing in a circle with my pals swilling cocktails and craft beers. I'm holding court, dazzling them with my oil field talk, blabbing. I tell them about Huck's bar fights and Porkchop's prison stories. I do my Olhouser impression. They're laughing, their eyes shining through the bleariness of the booze, excited by me, by the secret, distant life I bring them closer to.

"What is 'turtle fucking'?" my date asks.

"Good question!" I say. "I guess I don't know. I actually never got turtle fucked!" We Google it later to find out that the threat is scarier than the act. To turtle fuck someone is to slam your green hard hat on top of theirs. Now we know.

I'm soaking it up, this chance to be a raconteur, this attention I'm receiving for my work, this opportunity to tell the tale. What I do in Williston lands here, I realize, taking in the elegant woodwork and the chandeliers. New York City reaps the benefits of labor done thousands of miles away on the desolate plains of North Dakota, the labor I do. I feel proud. That is what it means to be a good hand: to do meaningful work.

And how rare is that? At this point in history, in the lives of most Americans? To spend a day doing work that matters?

It's a diamond.

But as for the result of that work? It appears to me like in a vision, this strange and startling fact: New York benefits from the oil boom far more than Williston ever will. No one here realizes that. Nobody even considers it. Here in this West Village gastropub. Look at them. Everything they enjoy. Every. Single. Thing. They get it from me. They get it from me and a group of the toughest, meanest motherfuckers I have met in my life. Men they wouldn't like, men they look down on, invisible men they will never see in a state they dismiss as flyover. They owe it all to the hands. All of it.

How is this? When we think of oil, mostly we think of it in terms of transportation—gas, diesel, and jet fuel. Refined oil and the petrochemicals it produces, however, are more invasive, more ubiquitous, and more life-altering than anything as simple as a truck or a plane.

In 1918, Fritz Haber won the Nobel Prize in Chemistry for creating a process, incorporating methane, to convert nitrogen—taken directly from the air—into ammonia. Ammonia extracted in this method has made possible the widespread use of nitrogen fertilizers. This is arguably the most significant scientific breakthrough in the history of mankind.

Before nitrogen fertilizers, humans relied on found organic materials such as compost, horse manure, and guano (bat shit) to fertilize crops. The Haber-Bosch Process changed all that. Nitrogen fertilizer has created the food resources that have allowed the world's population to balloon to 7.7 billion people today. Without refined methane, inseparable from the Haber-Bosch Process, most of us would have never been born. Most of our grandparents, if they were born at all, would have starved to death.

So, yes, New York City, you are welcome! I'm on a dance floor now, spinning my body in a tequila-fueled frenzy, pulling my date close up against me. The whole place is shaking, tilting like Atlantis. I laugh and stagger with the crowd. "You are welcome!" I shout, my voice drowning under the weight of the DJ's thumping bass.

"What did you say?" she puts her hand on my elbow and pulls her lips close to my ear. I pull away.

"You're welcome!" I shout it again to the room. She looks at me but still can't hear. She smiles a wicked smile and shrugs, then she closes her eyes, raises her arms above her head, and disappears into the music.

HUCK GETS BUSTED–AGAIN

It had been a month since Huck was last locked up. While I was in New York, I joked to my friends, "Hopefully, he won't be in jail when I get back to North Dakota."

When I get back to North Dakota, Huck is in jail. A few days later, we meet up for a couple drinks. This is the beginning of our friendship, the first time we get together outside of work, and Huck is as soft-spoken as Dr. Jekyll. "This is what happened, Magic," he says. His voice is flat, heavy, and listless. As usual, his troubles are tied to Krystal, and her boyfriend, Logan.

"We were at the Airport International Inn drinking Crown and Cokes, and I hadn't eaten all day. Krystal was trying to get Logan to leave the bar, but I'm like, 'You're being a fucking bitch.' She goes out to sit in the truck, and he stays to drink with me.

"How the cops ended up getting called this time was there was a fat lady singing karaoke and Logan didn't like it. She was putting her heart out there singing, and he was being disrespectful. Her friends said something to him, and I can see why. But I had to have Logan's back, so I started something with them 'cause they were starting something with Logan. And they're like, 'Let's step out-side.'"

"So you got in a fight with those guys," I say.

"The police report said there were people fighting outside, but I don't think I even got in a fight," Huck tells me. "When the cops showed up, I backed down because I've been through that rodeo before. I was like, 'You guys do what you gotta do, throw me on the ground, whatever, I'm not resisting at all.' But when they grabbed me, Logan freaked out. I think he punched one of them. They put him in the cop car and took off with him.

"Krystal's sitting there in her truck looking at me. I'm in the back of the cop car, and I start spinning. I got a big dip in my mouth, and the juice starts sliding back down my throat. I felt like I was gonna vomit, so I spit the dip out on the floor. And the cop is like, 'You fucking spit in my vehicle!'"

Huck shakes his head ruefully. I can't tell how much he is telling the truth versus how much he is practicing a version of events that need to be his truth in court. He continues, "He was about to let me go, but he took me to jail for criminal mischief. Fuck you, man! I should've puked in the back of his car. That's how that all hap-pened."

"Damn, dude," I say, leaning back in my chair, breathing in through my nose. "We gotta get you a jacket and tie."

"Yeah," he says, resigned.

"How long were you locked up?"

"Just twenty-four hours. They wouldn't let me in the courtroom for my arraignment this time, though. I was considered dangerous, so I was in my own room by myself looking at a TV and a camera. It was real dramatic."

I had run into Krystal and Logan's mother at the Williston

Municipal Court when I dropped in one day to pay a traffic ticket. They were trying to bail Logan out of jail. This was before Huck had told me the whole story. It was a tiny room, and I heard them mention Huck's name. I said, "I know him," and they stopped talking. Logan's mom appraised me coolly.

"He's not my favorite person," she said.

It was the only time I'd see Krystal in person. She was much younger than I imagined, birdlike and shy. We didn't exchange any words. I felt impolite for intruding on their conversation, and I quickly paid my ticket and left. I'd never meet Logan. I don't know if it was on purpose or not, but Huck kept Logan and me separate.

TRIPLES

Erwin "Jack" Jackson is tall and wiry, in his midforties, with short-cropped graying hair. He's got the low-slung broad shoulders and bandy arms of a gorilla coupled with the punchy bob-and-weave footwork of an amateur middleweight. I'm showing him how far to back the truck up to the edge of location, signaling the Wildebeest with my hands. Jack is keeping step with me. It's his first week of work. He asks how the job pays.

"Not as good as it should," I tell him. "But it's not bad. I'm working about seventy hours a week on average." The most I'd been paid at Diamondback was $3,039 for two weeks' work, but my average paycheck at this point was closer to $2,100. I was clearing just over a grand a week. It wasn't terrible, but it was far below what I thought I'd be making when I started.

Jack listens and nods. "Long as my child support doesn't catch up with me," he says.

"You've got kids, huh?"

"Yeeeaah. Nine kids with six different mamas."

"Jesus Christ," I say, "you need to learn to wrap your dick."

"Now you tell me!" he says, comically exasperated. I signal the

Beest to stop the truck and drop his chains. The winch releases and the load lowers into the mud.

"If I hadn't spent so much time in prison, I'd probably have twelve of 'em," Jack Jackson says cheerfully. "Yeeeeaaah, just finished eighteen months in Washington. Eighteen on twenty-seven. Before that I served five years in Florida. Before that fourteen months in New York State. Yeaahhh." He exhibits a sheepish pride telling me this, half-embarrassed and half-boastful.

"What were you in for?" I ask. We begin removing the shit hooks from the load.

"They locked me up for robbery in Washington, but that was bullshit. Anybody would tell you that. I'm not a fucking thief, you know. They just wanted to pin something on me. Yeaaaah, aggravated assault, mostly."

We get the steel unhooked, and I take a look at Jackson. He's big, with hands the size of frying pans. I start showing him how to re-hook the shit chains to get them out of the way when they aren't needed. He watches me, affable, relaxed.

I find myself drawn to Jack. He reminds me of someone I'd worshipped as a kid. Willie O'Shea was a wild redneck, reed thin with a jangly walk and a shock of unruly black hair like a hillbilly version of a young Nicolas Cage.

Willie was one of the only neighborhood kids who would drop by the farmhouse when I was growing up. My mom loved him; we treated him like family. He was funny and loyal as hell. Almost all of us have stories about Willie saving our asses. Once, while drag racing toward the local high school with my brother Matthew riding shotgun, Willie lost control of his Mustang. His immediate reaction was to use his right arm to pin my brother in his seat. Neither of them was wearing a seat belt. Willie held Matthew there while he got the car under control left-handed. Another time, when I was getting picked on by an older boy, Willie took the kid aside and told him that I was a martial arts knife expert. He must have been convincing because the older boy never picked on me again. Willie was like some kind of redneck guardian angel. When Kate was forced to work in the fields with my dad, he would sometimes show up and do the work for her. He was just like that. When Shanon died, Willie was a pallbearer. I'll never forget the sight of him dressed up with his hair combed down and a somber expression on his face. To my young brain, it was this image of Willie O'Shea in a suit and tie that brought home to me the fact that I would never see my sister again.

When he was eighteen, Willie got in trouble with the law. He'd been growing a field of marijuana in the woods not too far from where we lived, and he'd sold a little cocaine to a buddy. He was narked out by some scumbag who was working with the police to gain a lighter sentence for his own crimes. The police also got Willie for blowing up a pavilion in a local park. He liked doing dumb shit, liked making things go boom. They confiscated an AK-47 he'd won in a card game while he was in the National Guard, and the press painted him as some sort of white nationalist militia figure. He ended up doing eight years in Jessup.

He was big when he got out of prison, with a shaved head and a white power tattoo wrapped around a thick bicep. He didn't look like someone I'd be inclined to be friends with. But I accepted the tattoo as a reflection of his prison time and not his heart. To me, he was the same old Willie and I still love him like a brother.

I thought of him when I met Jack. Jack had the same gait, the same energy, and, as I'd learn, the same unfortunate tattoos. I'd end up forgiving Jack a lot of his transgressions because of the association I'd made between the two of them. The difference was that Willie never was violent. He wouldn't hurt a fly. Jack, however, would prove to be a very dangerous person.

"You don't seem like a fighter," I tell Jack, sizing him up for the first time, pausing to straighten a chain and slide a shit hook into a metal ring.

"I got a speech impediment," he says. "Whenever I open my mouth, somebody stands up."

I respond with a snort, and he continues. "Yeeeaah, man. Ask my friends, they'll tell you. They've never seen me start a fight. I just attract retards."

"Bad luck, huh?"

"I guess," he says, not totally agreeing.

Absentmindedly he raises a gloved hand to his mouth. His maw is a collection of busted, crooked, and missing teeth, the result of a stomp-down by Washington State's finest in retaliation for a beating Jack delivered to a sheriff's deputy he claimed was harassing him.

"They beat me down good. Yeah. Prison wouldn't pay for the dental work. It's why I look like a jack-o'-lantern," he tells me with that same combination of humility and pride. "Once I get some money in my pocket, though, I got this broad who works for a dentist. Yeeeaaaah. She's gonna fix me up."

With the hooks fastened correctly onto their chains, I signal the gin truck to boom up. The poles return to their original position, and the Beest gears off to the next pick. The voluble ex-con and I follow him over.

WHEN JACK JACKSON orders a double Crown and Coke, the waitress tells us the boomtown football night special: a triple shot of booze in your drink for the price of a double. It's the most ridiculous deal I've ever heard of in my life. I've never even heard of a "triple" outside of baseball. Why not, though? We order two very strong whiskey and Cokes, and by the end of the second round we're basically concussed.

"Yeah, I been shot two times," Jack tells me. "Still got one bullet in my foot. I was robbing this Mexican dude, he unloaded his whole fucking gun at me. Yeeeeaaah. Whole thing. Last bullet got my foot. Good thing he wasn't a good shot. I'm limping after him with a lead pipe, 'Get back here, you sonova . . .' Yeeeaaaah. He got away. I tried to get my buddy to cut it out, the bullet, ya know? I said, 'C'mon motherfucker!' but nah, he wouldn't do it. So, I got that lead in me."

We order a third round. "After this, I'm gonna need a stretcher to carry me outta here," I say.

"This hot little Puerto Rican broad that I had four kids with, ya know?" Jack says. "She's got a banging body. Still. Great tits. We lived together till she caught me fucking the babysitter. I had a kid with her later. The babysitter, yeah. But she tried to kill me, the Puerto Rican broad. When she caught us. Yeah. She came at me with a pair of scissors. I broke her nose, ya know. Yeaaaah. I call my kids with her my little half-breeds. I do pretty good for myself for looking like a jack-o'-lantern."

Jack Jackson has the charisma of a con man. He's a prison-hewn, rough-handed clown with a busted mouth and a story for every double (or triple) Crown and Coke he swills. Even as he talks candidly and proudly about his past crimes, they seem far away—silly situations he'd gotten himself into, like scenes in a movie. He seems easy, all goof, no scare. I am slightly wary of him, but I find myself fascinated by him, too. I encourage him to talk. And talk he does.

"LOTTA FUCKING CHOMOS around here," Jack tells me, using prison slang for child molesters. I've pulled into the Fox Run trailer park, and he's getting out of the vehicle. "Yeah, they're supposed to register with the police and then check in every thirty days, but they don't. So the cops run sweeps every month. This place is fucking crawling with 'em. Yeah. It really is." He stands outside the Blazer with his hand resting on the hood, looking around the trailer park like he might spot a chomo on the run. "Sure you don't wanna come in, Magic Mike?" he asks.

"Naw, I better get home," I tell him.

"Oh hey," he says in closing, "so I'm fucking this chick from behind, you know, and she calls me a pedophile!"

"Ugh," I say.

"Yeah, right?" he says. "I say, 'That's a pretty big word for a seven-year-old.'"

The sound that comes out of me is closer to a horror movie scream than a laugh. "Jesus Christ!" I shout. Jack watches my reaction with a shit-eating grin slapped across his face.

"Yeeeaaahhh. Anyway," he bobs his head and chuckles, walking toward the door of his fifth wheel, "I'll see ya later."

I FEEL THE triples slosh around my belly and my eyesight smears as I pull onto Route 2 and start heading south back toward town. It's a straight shot to my room and board. I just gotta keep 'er 'tween the ditches, as they say. As I'm driving, a tractor trailer frantically flashes its headlights at me. I must not have my headlights on, I think, or maybe my high beams are up. I play with the knob. No, my lights are fine. *What an asshole.* A moment later, I pass more traffic, another tractor trailer and a couple pickup trucks all flashing their headlights and blowing their horns. What's going on? A half second later, I realize I'm in the wrong lane. Holy fuck. I pulled onto the highway but didn't cross the median to the southbound side. I'm driving directly into northbound traffic. My heart starts banging through my chest. I swing around at the first opportunity and cross the median to the correct lane of traffic. Scared suddenly near sober, I drive slowly, carefully back into town.

YEAH BUDDY

Standing outside the Pilot blowing onto our coffees, it is Jack Jackson, Yeah Buddy, and me. Yeah Buddy, a crane operator with a syrupy southern accent and a giant head with eye bulbs as big as a Simpson's character, is talking:

"You know what the one thing I haven't had since I moved to this town is?"

"Eh?"

"A good cup of coffee. Yeah, buddy. This stuff tastes like hot water!" He kind of screams it before continuing, "You know one thing I've seen too much of?" He blinks the huge bulbs he has for eyes.

"Eh?"

"Ducks!"

He's right. Western North Dakota is smack dab in the middle of the Central Flyway, a migration route for birds traversing the Great Plains from as far south as Argentina to as far north as the Arctic Ocean. The place is stupid with them, mallards, pintails, puddle ducks, and Canadian geese.

"They're everywhere!" shouts Yeah Buddy.

Not long before, I had worked a rig move with a group of Native hands who every morning stopped on the way to the job site, rolled down the windows of the pickup, and, using slingshots with nuts and washers, shot at the baby ducks in the kettles beside the road. Mostly, they missed. But when they hit one they'd say, "He got sick."

Whenever I tried to talk, they told me to shut up and called me an out-of-towner. After a few mornings, though, they started giving me a hard time for being too quiet. "He's probably with the game warden," one of them said. We all had a pretty good laugh over that.

Yeah Buddy sips his coffee, his giant eyes glancing back and forth as if on the lookout for incoming ducks. Jack Jackson can't stand still. He bobs and weaves and shrugs about the ducks. Makes no difference to him.

We're moving the crane to a job site sixty miles east of town. The crane travels at forty-five miles per hour tops. At that rate we should reach the lease within a couple hours. Except this is Yeah Buddy. Due to his yapping and foot-dragging, it's already taken us over an hour just to get out of the yard. We've gone thirteen miles to the truck stop, and now we are taking a forty-minute break.

Because Yeah Buddy's gotta get a breakfast sandwich. "And spinach . . . Do y'all have any spinach? Yeah? Okay! Well . . . No, actually I'll just have ketchup."

Because Yeah Buddy's gotta get some coffee. "Man, remember them French vanilla little thingies, I really like them. Usually they're right here! Anyway, guess I'll . . . , huh, have ya ever tried hazelnut?"

Because Yeah Buddy's gotta pick out the perfect Danish. "Raspberries are pretty much just strawberries with little seeds in 'em."

And because Yeah Buddy's gotta make flirt-talk with the teenage girl at the food counter. "You got this gum in Spearmint? This is my brand; I just don't like Winter Cool. I'm cool enough, ain't I, honey?"

Finally, we get on the road, Yeah Buddy in the crane, Jack and I following behind in the pickup. Jack has been assigned Yeah Buddy's swamper; I'm on the job because Jack doesn't have a driver's license.

"DUI?" I ask.

Jack holds up six fingers.

"Jesus Christ," I say. "You don't do anything halfway, do you?"

"No." He shakes his head and grins. "I'd have a seventh one, but that one time I broke my neck, yeeeaaah, slammed my car into a wall at eighty miles an hour? Yeah, the cops figured I had enough problems already. They were laughing and waving the Breathalyzer at me as I was carted off on a gurney. Yeeeaaah. They didn't make me take it, though."

"That's nice of 'em."

"Yeeeaaahhhh," he agrees.

WE DRIVE AN HOUR, then stop at a second truck stop in Tioga to get more coffee and bullshit more. Then we drive another forty minutes and pull off again at the side of a county road for Yeah Buddy to stretch his legs. The crane is stressful to drive, and the ride is bumpy. "It'll jiggle up your insides," I'll hear him say later, but by just about any standard of hard work versus laziness, it is safe to say Yeah Buddy is stretching it.

THE MORNING I first met Yeah Buddy, I'd arrived at the yard to be told that a goddamn fucking swamper had called out, and I needed to take the goddamn fucking crane truck to the town of Ray to pick up that fat union fuck, Yeah Buddy.

"Yeah, buddy," said Yeah Buddy when I picked him up.

He was loud and gregarious, the first and only proud union man I met in North Dakota. He had been a helicopter mechanic in the army but injured his leg in a crash and became a crane operator in Atlanta after seeing an ad in the paper, walking to a job site and being told by the foreman to hop in and give it a whirl. It was a different time. There was more construction happening in downtown Atlanta than there were construction workers. "Me and the union built that city," Yeah Buddy told me. After that, he moved to Massachusetts to work the Big Dig. The most expensive highway project in US history.

"Same job site for five years," he said. "Yeah, buddy."

I liked him. Having lived in Boston, Yeah Buddy could talk about the East Coast and he was happy to banter about something other than work. We had a fine conversation that first morning. We hit it off, I felt, me and Yeah Buddy.

Once we got to the work site, however, his easygoing demeanor devolved into finger wagging and narrow-eyed impatience. "I'm only going to show you how to do this one time," Yeah Buddy said, and then "Gosh dangit!" he'd holler if I got it wrong.

His frustration was more comical than intimidating, but that didn't make it any less annoying. And it didn't help that he played into the conservative stereotype of a lazy union man, taking longer breaks than the others, talking slower, driving slower, and repeatedly asking me to grab him soda from the cooler in the truck.

I nearly fell out of my seat when, driving him home at the end of the day, he resumed our conversation from the morning in the exact same easygoing, laid-back way that it had started. Before I dropped him off, he even offered to buy me dinner. After fourteen hours of listening to him jaw, I declined. But ever since that day, I'd vacillated between getting a kick out of Yeah Buddy and detesting him.

"Just wondering where the fuck anybody is down here on 138," Huck's voice comes over the company radio. "We're waiting for you. Still, motherfuckers!" The radio is in and out, and at first Jack and I aren't able to respond. Huck continues, "This ain't a fucking union job, buddy. We're ready to get working . . ." When we finally do get through to him, Huck snaps at me and calls Yeah Buddy a slow-moving dumbass.

When we pull onto the lease, the irritation is palpable. A row of stone-faced oil field hands greet us by not greeting us. Jack and I shrug and hop out of the truck.

There are only a few trucks on location and maybe six hands. Every one of them watches Yeah Buddy with well-practiced expressions of disappointment and disgust. Every one of them is as handy with a withering stare as with a sledgehammer. And these boys can swing. Yeah Buddy climbs out of the cab with his hard hat on his head and his safety glasses askew, his britches slipping down his ass, and he starts barking orders. He doesn't work well when he's got the screws put to him, and right now he's got the screws put to him. The men stare at him, like mafiosos confronted by Barney Fife. Nobody budges.

Yeah Buddy swings his head around until his eyes land on me. "Go and get that truck with the trailer on it and swing it back around over here."

Jack and I exchange a look, and I head in the direction of the trailer. My movement breaks the spell of stillness and the other hands start moving, too. I hop in the truck and begin to pull the rig around, but I've barely ever driven a trailer. I am not very good at it. Turning right to get left is a trick I can't quite master in the moment. Soon Yeah Buddy is hollering and his arms are flailing at me to stop. Huck and a couple amused hands look on.

"Well, I've never done it before and I'm trying to learn," I tell Yeah Buddy.

"Gosh Almighty, I said 'here!'" he says. "If you can't do it . . ."

"I mean, you want me to learn, right? If I practice," I tell him, "then I'll be able to do it later on."

"Yeah," says Huck in a slow, mocking tone, "you want him to learn, don't you?"

"Gosh dangit, go ahead! Just get it over here," says Yeah Buddy.

I give it another try, and completely miss the mark again. I am feeling the heat but only a little bit. Mostly I'm enjoying the apoplectic response of Yeah Buddy. The other hands can see this, so they don't mind that I'm slowing things down.

"Gosh dangit!" Yeah Buddy shouts. He orders me out of the truck.

"Give a man a fish . . ." I say. Huck hops in the driver's seat, and within ten seconds the truck and trailer are perfectly positioned.

A FEW HOURS LATER, I'm tasked to drive us back to the yard. Yeah Buddy sits up front, and Huck folds himself up in the back seat. Next to Huck is a silent, visibly stewing Jack Jackson.

As we rigged up the crane, Yeah Buddy started nagging Jack. He talked down to him, and Jack clammed up. His face went stone. His expression stopped. The bandy muscles that wrapped his shoulders and arms began flexing, and his wrecked-mouth smile disappeared into a thin, grim line. I've seen men grow angry before, but I'd never seen anything like this. The small provocation set off a seismic shift in Jack's very being. It was like a glitch in his core processor, and it turned him into a very different man.

Over the past week, I'd laughed off Jack's stories of brawls and gunfights, believing but dismissing them as a combination of macho exaggeration mixed with the bad boy nostalgia of the reformed. On this day, I got a sense of just what those aggravated assault charges must have looked like to the victims. Later, Jack would tell me he once beat a guy so senseless he gave him brain damage. "His family hates me, you know. I made him a, yeeeeaaaah, a retard." This side of Jack was terrifying. The switch in his personality was not night and day; it was life and death. He removed his hard hat and threw it across the dirt, then he stomped after it.

"What's with him?" Huck asked me, surprised.

"I dunno," I said quietly.

The new Jack looked like he could punch his way through a bulldozer. I kept my distance from him and he cooled as we worked, but still, even on the way home, I could sense his simmer in the back seat behind me.

I weave the Dodge down the dirt oil field roads and onto a county highway.

We pass pastures of canola and a field full of bright yellow sunflowers. We catch sight of several hawks, swooping and drifting against a momentous sky. We pass a cemetery.

"You know how many dead people are in that cemetery?" Huck asks from the back seat.

"No."

"All of 'em."

"You don't say?" I tell him.

"Oh yeah." He smiles, revealing the dip lodged in his lower lip. "You know why they got gates around cemeteries?"

"No."

"Because everybody is dying to get in."

WE GET STUCK behind a pickup truck on a long straight stretch of highway. I'm maintaining speed, and that starts to drive everybody crazy. They want me to pass it. I'm not used to driving Yeah Buddy's Dodge pickup, and I feel like we're going fast enough, but I cave to the peer pressure and push the Dodge around the truck in front of us. We start to speed up as I'm getting around the truck when the hill begins sloping upward. I have to lean pretty heavy on the gas pedal. At around seventy miles an hour, the Dodge starts to shake. I keep pushing. The wheel is rattling in my hands when the Dodge begins to wobble on its wheels. Seventy-five. I feel like it could all come apart on me. Eighty. I grit my teeth and keep my eyes on the road, finally making it around the truck and breathing out in relief.

"If we'd have been in that other crane pickup, you would've just wrecked us," Yeah Buddy chimes in.

"Fuck off," I say.

I'm stiff from driving. I pull off onto the side of the road to take a break. If I could, I'd pass the driving duties to somebody else, but Yeah Buddy would never do it, and neither Huck nor Jack has a valid license. We get out of the truck and stand in a wide, splendid field. The sun sits low on the horizon, throwing a yellow light sideways across the high plains. Butterflies flit around. Huck inspects the crop we're standing in.

"This is durum wheat," he claims. "They use it in making macaroni and gum." He takes a piece of it in his mouth and chews it gently, thoughtfully. "You chew it and you can feel it growing kinda gummy in your mouth," he says. He spits it out.

The whole world swirls around me, around us. I feel my consciousness twist and expand and grow. The light plays with the wheat. A butterfly lands on my hand. It is almost as if I have taken psychedelics. The graveyards we passed seem to follow us to this point, this razor's edge, a sense of life as thin as tissue paper, threatening to tear under even the gentle strain of breath. I think of staying in Williston through the winter. The feeling is unnerving, sickly sweet in my belly and liberating.

FROG

You'd better kick it down, or we're gonna be late," Frog says impatiently from the swamp seat. I mumble something about not seeing well in the dark. It is early October, the days are shrinking, and morning is pitch black before sunrise. "Olhouser'll raise hell," Frog says.

I kick down on the gas pedal, and the pickup opens up. The asphalt traveling west from North Dakota turns to gravel at the Montana line and then drops off into a spare one-lane dirt road. We jostle our way through the winding nothing, traveling mostly in silence, Frog falling in and out of sleep as I blink my eyes wide and push it faster than I want to go. I keep my focus on the quick turns and sloping bends of the road.

"Is this our turn coming up?" I ask.

Frog doesn't open his eyes. "No, it's further down," he says.

"I guess I'm used to landmarks," I drawl.

"Out here you have to read the blades of grass," he says. Fred "Frog" Eagle is Chippewa. "If it weren't for you white fuckers showing up," he is fond of saying, "I'd be hunting and fishing all day long."

It's a good quip, and he means it. He's a real gamesman. When the crane is quiet, Frog sits in the cab, his "Native Pride" baseball cap perched high on his head, flipping through hunting magazines.

"Stop reading and get the fuck back to work!" a roughneck once yelled at him.

"Reading?" Frog responded. "I'm looking at the pictures."

I bring us onto location just in time for the morning's safety meeting. Frog slaps his hard hat onto his head and hops out. I suit up in the front seat. With the turn to autumn the weather has gone from cool to cold in a twig snap. Pulling on layers of clothes, I climb out of the truck and walk through the chilly bite of predawn morning and into the driller's shack.

The shack is full. Roughnecks have taken up all the seating. Rig up hands line the walls. I see Huck step in sleepy-eyed. The cold blackness of day pushes its way in through the door. You can feel its fingers when it touches you. Huck takes a quick look around and slips back out, no doubt to catch a few more z's in the gin truck. Someone hollers at Tex to shut the fuckin' door, and he does. Frog squeezes into a seat and trades bullshit with a group of Canadian floor hands.

Olhouser is in high spirits, frenetically moving about the shack, jostling and joking with the guys, cussing things he can't stand: southerners, fat fucks, and wormy operators. Cheerfully, he tells the driller how he ran off Yeah Buddy. "I told him to get his fat ass back to Mississippi!" he cackles. Yeah Buddy is from Georgia. *Oh, well*, I think.

"He's been running his swampers around for pop!" Olhouser says. His mole-like eyes dart around the room until they settle on me. "He ever run you around for pop? That's what that other guy said."

I shrug. I hadn't swamped for Yeah Buddy in a long time. But he was known to be a pain in the ass, always sending his swampers on meaningless, menial tasks. By the time they paired him with Jack Jackson, he'd already worn out a few guys.

"Bobby Lee, he tells me I need to calm down," Olhouser announces to the room. "'You've got to change! It's a new oil field,' he says. Well, I'm sixty-three years old and I been doing this since I was knee-high to a grasshopper." Olhouser's spirited voice rises higher in pitch into a rousing Mickey Mouse squeal: "I'm not changing a fucking thing!"

We laugh, nod, and glance around in agreement, grinning when we catch each other's bemused and crusty eyes. I feel proud of the old man, proud to be working under him. Maybe it isn't ridiculous to be inspired, I think. Inspired by work. Olhouser is the last of his kind in this small world of field workers, but it is bigger than that: he is the last of the good hands, the rough, calloused hands, cracked and hardened by the vagaries of the work that makes this world spin. Bobby Olhouser is Pecos Bill and Paul Bunyan stuffed into one leathered hide, bathed in Bakken black gold, drenched in the cold kettles of spring, coated by summer's heated dust storms and blinded in this Northland's winter whiteouts. He is true; an oil field legend. We are glad to be there with him.

DRIVING BACK AND FORTH to the job, I keep losing my way. "I used to have a good sense of direction," I tell Frog, "but I moved out here and my radar broke."

"Read the blades of grass, Magic," he tells me.

I learn to read the blades of grass better, but not much better. One evening leaving location, I pull down the wrong road and onto a second patch of scoria. I drive in an idiot's circle around a pump jack before realizing my mistake.

"Wrong blade," says Frog.

By the third day, I know the route. Sipping coffee, winding the pickup through the dry creek beds that pass for roads in this part of Eastern Montana, I rub the sleep from my eyes as Frog snores contentedly next to me.

Frog can't wait for winter to set in, to really set in, "For that cold Canadian wind to blow all the fucking out-of-towners back to where they come from." Scatter the idiots to all corners, the Floridians, the Mississippians, the Idahoans, and the Texans, "for God's sake at least the fucking Texans."

I don't take it personally, his out-of-towner talk, not at this point. If I did, I wouldn't have lasted a boomtown day. Being from New York, I may as well be from the farther reaches of outer space. I understand Frog's frustration better when he tells me of a friend of his who speaks fluent Crow. "She works at Walmart in the checkout lane. And she was talking on the phone when some fucking out-of-towner says, 'If you can't speak the language, then you better go back to wherever the fuck you come from.' She said, 'My people have been here a thousand years longer than yours, asshole.' That shut him up."

FROG HAS A pair of buffalo hides in the back of his pickup, and Huck offers to buy them for $50. "If they're cleaned up properly and treated, you can sell them for a couple thousand dollars," Huck tells me.

Frog tells Huck he can have them for free. "They're a little fucked up," he says.

When we finish the rig up, Huck hops in my swamp seat, and we bounce off down the dirt road with Frog trailing us in the rigged-down crane. It is daylight and I want to get home, so I push into it and keep her pinned. The road hugs the prairie's curves. Gravel and dirt spit out behind us. But we get too far in front of the crane, and Frog's voice comes barking over the radio. "Slow the fuck down."

I slow down, chewing the fat with Huck. He is in hang-dog mode, a basset hound in the swamp seat telling me his troubles: police troubles, lawyer troubles, friend troubles, and love troubles. I nod along, half listening, when I realize I have slowed down too much. The crane is less than five yards behind us. In the rearview mirror, I see Frog leaning forward in the cab with a clenched smile wrapped tightly around his face, bearing down fast. I hit the gas, but the truck doesn't respond very quickly.

I've got the pedal pinned to the floor, the wheels spinning in the dirt at the bottom of a brief, low curve in the valley. The crane is directly on top of us. Frog all but standing on the gas, he's pressed against the windshield, his camo cap perched on his head, eyes glowing with an unbridled lust for speed. The crane swallows all my mirrors. I feel it inching even closer—the back of the pickup beginning to shudder. My foot pumps the pedal, and fuel spits into the engine. The piston travels down on the intake stroke, the valve compresses the fuel, a spark plug sparks, and combustion provides power. The Dodge finds its ass. Tires dig into earth. We push out and up and away, slowly stretching the distance between the tail-pipe and the bumper of the crane. Once we've burned farther down the dirt road, I'm sure to maintain the proper distance between the pickup and crane. Lesson learned.

Huck barely notices any of it. He sits, his knees pushed up against the dash, sipping on a cigarette with the window cracked, talking about the woman he loves.

I fight a smile. It is comical. Huck! Who delights in being a field hand, spitting dip, making trouble, and cussing. Huck! Who loves being loud and brash, bigger than life. Huck! Who *is* bigger than life. Huck has it so bad for Krystal, and he can't have her. He might punch Logan in the head, but he's not going to sleep with his girl-friend. He tells me about riding through town with Krystal, about a text she sent, a comment she made, a passing glance he caught, and I see the waves of elation and despair course through his body, filling and emptying him of hope and heartache—sometimes in the course of a single sentence. He is deep in the throes of a crush that is closer to pneumonia than love. "I think about her more than I used to think about fly-fishing," he tells me earnestly. "She's all I think about." He can't get her out of his head or his heart or his loins. She's all he jerks off to.

"I don't actually like to fight," Huck tells me. I scoff. "I really don't! I don't feel like a badass. It's just because I'm so big, guys always wanna start shit with me. But I don't feel big," he says. "I feel small."

"I'm small, and I feel big," I tell him. As a sick kid I always felt scrawny, and that feeling returned when I got to Williston because I didn't know anyone, I didn't know how to do anything, and most of the people I met were physically larger than me. Before that, in New York, walking out onto a stage with a guitar in my hand, I could feel huge. And now that I've gotten good at oil field work, I feel big again. "Maybe we were switched," I crack.

"Maybe, Magic," Huck says. "Maybe."

Talking about living leads to talking about being born, and soon we are talking about our fathers. Huck's dad has a mean streak. "He punched me through a door one time," Huck tells me, and then repeats it for emphasis. "I was standing on the other side of the door, and he punched through the fucking door to hit me."

I think of all the men I've met who bonded over this particular tear in their souls, and I'm silent watching the prairie pass. Huck laughs it off. He asks about my dad, and I keep looking out the window. I guess some men talk about golf, I think. Some men talk about music. Or college. But not here where the oil brought a boom of beaten sons. The grease may turn the world, but the wound rules the men who do the work.

I tell Huck that when I was in college my grandmother passed away. My dad and I had reconciled after a period of estrangement, and he enlisted me to help him clean out and repair his mother's house in order that he might sell it. The house was a single-story trailer that sat a hundred or so feet from a deep gullied river in West Virginia, just across the line from Cumberland, Maryland.

My father wanted to carpet one of the odd-shaped rooms, so I grabbed a tape measure and began to tally the size. "Oh no-no." My dad stopped me. "We don't need to do that," he said. "I'll just pace it off."

And pace it off he did. Setting one foot in front of the other toe to heel, counting each step. "Dad, I have a tape measure. Why don't we just measure it so you know how much carpet to buy?"

"I'll compromise with you," he told me. "You measure that wall and I'll pace the rest."

When the carpet store explained they couldn't cut a carpet based on the size of his shoe, he bought more than we needed and we spent hours trying to unfurl it in the room and cut it as we went. It took all afternoon and the angles we made by hand were awkward and unskilled.

"Looks pretty good," my father said, proudly staring at the finished job. It did not look good.

"The thing is," I say to Huck, "my old man had one of the gnarliest childhoods I've ever even read about, and he never complained about it. His dad was a piss-the-bed drunk, and his own mother abandoned him to an orphanage when he was eight or nine. She said she'd broken her ankle.

"He was in the orphanage for two years. The place was full of sexual predators, and run by a couple called Ma and Pa. Pa would make the boys walk in a circle with himself at the center and he would whip their bare backs with a leather belt. Ma handled the girls.

"He started boxing when he was a teenager," I tell Huck. "To hear him tell it, he was never any good, but my uncle Bill told me Dad used to whip everybody's ass."

He'd been unemployed when he volunteered to fight in Korea. "Bill, who was a fucking sea captain, told me Dad hit some of the 'hairiest goddamn fighting of that war.' He served for three years and received two Bronze Stars, but he didn't talk much about that, either. 'I think they gave them out to everybody,' he once said. They don't." I pause and glance over at Huck as I drive. I'm not sure why—I rarely talk about my dad, but I'm finding it easy somehow to unburden myself to a younger guy who is struggling with his own father issues.

"Occasionally he'd let a few things slip," I continue. "He talked sometimes about a woman and child he'd seen killed by napalm. I know it haunted him. And once he was ordered to murder a group of prisoners. 'I thought the CO was joking,' Dad told me. 'I laughed at him, so he ordered the guy next to me to do it. That guy shot them all in the head.'"

"This doesn't make the things my dad has done right," I say, "but if I don't view him through the lens of his own life . . . how fucking ungrateful can you be?"

· · ·

MY FATHER WOULD pass away in late 2019. In what I'd view as an incredible act of compassion, each one of my siblings came to be by his side. He wouldn't even be nice to all of us! Even there, camped out on death's door, he'd act like a dick. But over a period of a few weeks, Dad would grow nonverbal. Ryan, Matthew, Megan, Kate, and I, attended to, I am certain, by the benevolent spirit of our late, eldest sister Shanon, would sit by his side for hours on end; we'd wrap our arms around Dad, and we'd tell him we loved him.

My father's tragedy, I think, was his inability to accept love. There, in a hospital bed, bullied mute and laid prostrate by the ravages of time, he was forced to watch and to listen only. My brothers and sisters and I could have said anything we wanted to him. What we said was love.

One evening right near the end, several days after he had stopped talking, Dad popped up in bed and chatted for a moment with Matthew. Then he stared off into the distance. "What you thinking about, Pops?" Matthew asked him.

"My temper," he said. They would prove to be his last words.

"NOBODY ASKS ANYBODY to change blood types, do they?" I tell Huck. "My old man pushed the wheel forward. But that doesn't mean that he never scared the hell out of me, either."

I remember him driving me back from my grandmother's house the night after we did the carpet, and he began to doze off at the wheel. I was so terrified by the man and his temper, even when we were getting along, that I said nothing to wake him. Instead, I watched in a slow-motion sort of horror as, behind the wheel, his eyes would close, his head would start to droop, and when the car hit the rumble strips on the side of the road, he would jerk awake. I was frozen, seized with a complete and total fear, so scared of my own dad that I was willing to get into a car wreck to avoid the slightest chance of provoking his anger.

"It was a long time ago for me," I tell Huck. "It isn't so much that I'm over it. Just that time changes it. You're still so close to it."

"Yeah," says Huck.

"Besides that," I say, "your old man sounds a lot rougher than mine."

We reach the state line and, as North Dakota law doesn't require a flag-car escort for the crane, we break free of Fred "Frog" Eagle, his Native Pride cap, and his machine. We speed back toward the yard.

THE HIDES

We smell them first, slamming into a noxious wall a few feet from Frog's truck. Then, stepping up to the tailgate, standing in the stink of rotting flesh and animal shit, we see them. They lie in a fleshy heap in the bed of the truck, still full of meat and covered in excrement: the buffalo hides. They'd been in Frog's truck for four days. The flies buzzing over the hides had spawned maggots, curling and digging, wiggling in the remains. Huck pulls his shirt over his nose. I clench my stomach and breathe shallow out my mouth.

Until recently, the skins were wrapped tight around the warm, beating-hearted, full-blooded bodies of a young bull and a year-old heifer. "The bull," Frog tells us, "accidentally gored the yearling through its shit box." The yearling died in what Frog describes as a comedic dance of gore, spraying its own blood and shit out of the hole in its side.

"I'm not sure I want these," says Huck.

Frog tells Huck that, if he does want them, he should take the skins to the car wash and use the power washer on them. "That will get rid of the shit, the blood, and the maggots," he explains. Huck gathers his gumption. He pulls his Chevy around, throws the hides into the bed, and bounces off to the car wash, trailing the cheerful stink of death and shit.

IT IS NIGHTTIME when I pull off into a small parking lot against the Little Muddy River. The river spills under the 1804 bridge, sparkling and crackling in a deep, mud-walled gully that runs past the edge of the dirt lot.

Huck's pickup faces east. The buffalo hides are draped over the tailgate. Smash has pulled his truck around to face them, headlights illuminating the skins. I park my SUV, cut the engine, and walk toward them. I see the figure of Huck moving nimbly in and out of the headlights' glare, throwing shadows against the night, a long, curved knife in his hand.

The shit stink is gone—the power wash did its job—but the sweet smell of death remains. Huck carefully carves several small chunks of flesh from the hide. The meat drops to the gravel. Flies and mosquitoes buzz the skins. He steps out of the headlights' beam and into the surrounding darkness and sharpens his blade on a leather strap. The buffalo meat is tough, and his blade grows dull every few cuts.

They'd been drinking beer for a while, he and Smash. We converse lazily, our breath producing puffs of clouds in front of us, in the cold, clear North Dakota night, cracking jokes about the workday, downing beers, and tossing the empties into the bed of Smash's rig.

Huck settles into a rhythm with the work. It is entrancing to watch him, illuminated by headlights as if by fire, slicing into the meat against the skin, then stepping back from it, into the darkness and swiping the blade all in one long fluid motion down the length of the leather pelt, then back into the light—a slow, ancient dance.

When Huck is satisfied, he pulls the skins into his truck bed. Leaving his Chevy by the Little Muddy, he hops in with Smash, and I follow them into town to a bar called Doc Holliday's.

SMASH IS SICK of the same old drinks. Beer and whiskey, rum and Coke, vodka and orange juice. He wants something different. He wants a cocktail. He wants a Manhattan! It's one of my favorite drinks, so I order them just how I like them, on the rocks with bourbon in lieu of rye. The first sip is bitter, sweet, and delicious.

I think of New York. When I first moved there, I met a young woman drinking Manhattans at a Hell's Kitchen dive bar. We ditched our friends and shared a cab downtown, spent the night together on an inflatable air mattress in an empty East Village apartment. I don't know if I ever knew her name. I would never speak to her again, but that night was something of a perfect moment, carnal, urban, sweet like vermouth, and guilt-free. I tell the boys the story. The morning after, she offered me tea. I accepted, sitting at a high-top table in a postage-stamp-sized kitchen. By the time the tea was brewed, it was clear we didn't know what to say to each other, and I left after just a few sips.

Smash talks about how much he wants to have sex with guys. "I'm the gayest straight man you'll ever meet," he says. We have had a few drinks at that point, and Huck just shrugs. I'm not sure how much Smash is joking, if at all. Most field hands make a show of how straight and macho they are, but for all the homophobic language I sometimes hear in the field, almost nobody gives a shit who is gay or straight. At one point over the summer, a couple guys had seen photos on a swamper's phone that showed him giving another man a blow job. It was talked about for maybe a day. I don't think anyone even gave the kid a hard time. Later, I'd work with Charley, a crane operator, and some hands, while vociferously proclaiming they would never do it, would ask us if we would suck a dick for a million dollars—a great philosophical question! "A million dollars?" Charley would respond, completely disturbed. "Are you kidding? I'd suck a dick for half that! A tenth of that! I'd do a good job, too. I'd work the head! I'd massage the balls! What are you guys, a couple faggots?"

Huck, wearing a T-shirt that says "I Support the Right to Arm," orders us Peruvian Bear Fuckers, beer mixed with orange juice and a shot of Captain Morgan. They taste like Creamsicles. We slam them and get more beers to wash the sweet taste out of our mouths.

Someone takes photos of us. I'm standing in the middle in both pictures. In the first photo, I'm dwarfed by Huck and Smash. In the second one, I wrap my arms around the boys and they lift me up off my feet, so it almost looks like we are the same height.

I'd been so damn lonely all summer long. From June through September, my solitude ate at me like the heat; it wore me down every bit as much as the taunts of bed truck operators. Now it is cold and getting colder. I've proven myself on the job. I'm a decent hand. And, for the first time since arriving, I've made friends. I had no idea how important that would be to me. It is shocking to feel such a stark need I didn't even know I had suddenly filled. Also, I've decided to stay in town through winter—I can't shirk from a little cold. I ain't gonna let it whip me! I need to find out what all the fuss is about, anyway. Even in June, July, and August, North Dakotans love to talk about how cold it gets. I'm ready for it. My heart is full of joy.

THE NEXT MORNING, the entire side of his face is bloated and malformed, his right eye a bloodshot slit looking out from a puffy pool of purple flesh. The skin stretched almost translucent, pink, red, and green, cradles that purple eye. Only after looking at the left side of his face can I even see that it's Huck: the arched eyebrow; the long, pointed chin; the plaintive, sad eye; and the barely discernible smile.

When we parted ways, Huck and Smash had driven across the street to a local honky-tonk to play a game of pool.

"This big guy came in with a group of friends," Huck later tells me. "He was being a smart-ass, talking shit all night. When we got to the parking lot, I was like, 'I'm invincible. Fuck you! You can't hurt me, bitch.' Then he proceeded to beat the shit out of me. He's kind of a pussy, though. Somebody that drunk? My eyes were rolling in the back of my head. I feel bad when I fight people that are any kind of drunk."

Smash ran at the big guy to get him off Huck, but the big guy laid him out with a single punch. When he moved toward Smash to stomp him, his own buddies intervened, holding him back, which allowed Smash to crawl onto his knees and nurse his jaw. The group scrambled away before either Huck or Smash were back on their feet.

Huck was chastened, but Smash was pissed. The shame of seeing his friend beaten enraged him. He stormed through the scrap of dirt behind the honky-tonk, punching and kicking the parked trucks. An old cut on his forearm split open and blood stained his arm black before Huck could calm him down. They were lucky no one walked outside during his tirade. They would have had another fight, or a group of cops, to deal with.

Smash drove Huck home, but the incident burned at him. He went back to his apartment and grabbed his shotgun. He drove around all night looking for the big guy. He was going to scare him. Or kill him. But he never found him.

"You think he really would have shot the guy?" I'd later ask Huck.

"He's psycho enough," Huck says. "He doesn't like fighting but when you get him to that point he's like a murderous psycho. He's probably a fucking lunatic."

That afternoon, Huck heads back to the parking lot off the Little Muddy River to pick up his truck and get the buffalo hides. His truck is there, but the hides are gone, either stolen or dragged off by some animal.

BUFFALO SKINNERS

When I first arrived in Williston, I spent over a week camping in Theodore Roosevelt National Park south of town. Once, returning from a day of job hunting, I drove into the park at night. Rain was coming down in sheets—the end of the muddy spring of my arrival. I drove carefully, cresting one of the twisty hills leading back to my camp, when a huge, dark shape appeared on the road in front of me. I slowed to a stop and found myself face-to-face with an enormous bison. The Blazer's headlights reflecting through the rain gave the beast an unholy glow. It stood with its thick shoulders hunched, its black eyes buried deep in its massive head. It turned toward me and reckoned me, staring it seemed with the rage of the useless—a slaughtered species, a herd-less herd animal. In the bison, I saw not just a reminder of the plight of the Natives who honored them but also a reflection of the men who drove them to near extinction.

To the Native Americans of the plains, buffalo were as important to their way of life as oil now is to modern people. The difference is that the Indians lived what the Sioux called *wouncage*, or the Sacred Trust, what Paul VanDevelder in his book *Coyote Warrior* calls "the enlightened state of living in harmony and balance with the 'great mysterious.'"

Native Americans tailored buffalo hides into clothing. Bones were fashioned into knives and shovels and sleds. Hair was stuffed into pillows, sinew turned to thread; stomachs, bladders, and intestines made into waterproof containers. The horns were carved into jewelry, spoons, and cups; hoofs made into glue. When Natives of the Plains were laid to death on burial sites of raised scaffolding, they were often wrapped in fresh, still-moist buffalo skin.

As tribes across the plains, including the Sioux, would sing:

The buffaloes I, the buffaloes I,
I make the buffaloes march around,
I am related to the buffaloes, the buffaloes.

The white man sang a different tune. The greatest cowboy song ever sung is a dirty, grime-soaked ballad called "On the Trail to Mexico," or, more simply, "Buffalo Skinners." Like many canonical folk songs, there are multiple versions containing varied verses. Woody Guthrie recorded this song, but the best rendition you'll ever hear is sung by a protégé of Woody's, the flat-picking cowboy poet Ramblin' Jack Elliott.

A true apostle of American music, Ramblin' Jack was born Elliott Adnopoz, the son of a Jewish surgeon in Brooklyn, New York. He ran away from home as a teenager and joined the rodeo out of Madison Square Garden, mucking stalls and learning guitar from a cowboy and a rodeo clown. Jack was embarrassed at having been born in New York City, but the rodeo clown assured him: "It's not where you're from that matters. It's where you're going."

I've opened up shows for Jack Elliott a few times, and I've learned more about life and music sitting backstage talking nonsense with him than I've learned listening to a thousand records. "Most my friends are dead," he told me the first time we met. I could see the ghosts hanging off him. He'd been a compatriot and muse to some of the greatest writers and songwriters of the twentieth century. He was dripping with mischievous spirits and long, winding stories. Whether telling me about the difficulty of obtaining a working bugle from a pawn shop, or expressing regret at stealing a bull rider's girlfriend, or explaining the trials of teaching Allen Ginsberg how to tie a bowline knot, horses ran through his heart and oceans glittered in his eyes. His life seemed defined by his adventures, his friendships, and his ability to talk about them, more so even than the musical tradition he'd been carrying on for nearly seventy years. Yet, when he sings "The Night Herding Song," you can all but smell the campfire.

Jack's life upended any idea I might have had about the word *authenticity*. If you are born in the wrong place, at the wrong time, to the wrong family, you have to travel a lot of miles to get home. Earning it might just be more authentic than being born into it. The Jewish surgeon's son used the mud of his own body to fashion his very self into the thing only he knew he was. Ramblin' Jack Elliott became an Actual Fuckin' Cowboy. His version of "Buffalo Skinners" is killer, too.

In the song, the narrator, a young cowboy, hires on with a mysterious man named Crego. In Jack's rendition Crego is simply called "the drover." The drover tells the narrator:

> But if you do grow homesick and try to run away
> You'll starve to death out on the trail
> And also lose your pay

Living in North Dakota, I felt incredibly close to this ballad, I listened to recordings, and I played it over and over again on the guitar. In some ways, I felt like I was living inside of it.

In the song, the cowboy takes the drover up on his offer and sets out on the trail. A lightning storm hits. In some renditions, the cowboy falls off his horse and into a cactus. He gets *all full of stickers*! In another, he nearly cuts off his thumb. Hardships pile up, but onward the men travel, *skinning the damned old stinkers*, becoming the very thing they hunt. As an early rendition, collected in 1911 by John Lomax, laments:

> *Our hearts were cased in buffalo hocks, our souls were*
> > *cased in steel,*
> *The hardship of that summer would nearly make us reel,*

Like modern society wastes the petroleum that makes our lives possible, these cowboys killed buffalo indiscriminately, tricking the herd to race in a circle and then shooting into the mass of moving animals. They showed the bison no respect, treating them like we treat empty water bottles or unfinished plates of food. The song is rooted in real history.

At the end of the ballad, as Ramblin' Jack sings it, the drover refuses to pay the men:

> *But the cowboys never had heard of such a thing as a*
> > *bankrupt law,*
> *So we left that drover's bones to bleach in the hot,*
> > *desert sun.*

The drover is given the same disrespectful send-off that the men give to the animals they were paid to kill. It is a fun, B-movie type of ending that romanticizes frontier justice, but historically, the whole ordeal was a suicide pact. Cowboys killed off the buffalo at the behest of railroad magnates, and when they fed the armies that murdered and imprisoned America's original inhabitants, they destroyed the West and broke their own way of life. They did it for some hardtack, whiskey, and maybe a little bit of money. The result isn't so different from our society's addiction to fossil fuels.

And so I'd pick out "Buffalo Skinners" every other night on my guitar. And I'd see the faces of the cowboys matching with the faces of the field hands I knew and loved, small tragic figures under great big skies, foot soldiers in a war for progress that benefits the world even as it strangles it.

In Theodore Roosevelt National Park, in the pouring rain, illuminated in my headlights I stared directly into the dark eyes of a buffalo. Like the cowboys in "Buffalo Skinners" and the roughnecks of Williston, the animal before me exuded a white-hot anger and a deep-seated, entrenched sadness. I felt it boiling and roiling under the creature's lonesome skin, the skin of the marginalized, the unemployable, the passed over, the forgotten.

> *The buffaloes I, the buffaloes I*
> *I make the buffaloes march around,*
> *I am related to the buffaloes, the buffaloes.*

I moved the SUV around the animal, fearful it would charge. I then watched it through the rearview, receding silently from sight in the red glow of my taillights.

WOLFISH

I walk into J Dub's trailing a cloud of dust and grab a stool at the bar. I'm wearing greasy jeans and a dirty T-shirt, my beard is overgrown, and my face is a dark reddish-brown below the line where safety glasses fit my face. My dirty John Deere cap covers my matted, sweaty hair. I've come straight from work, and I'm wolfish with hunger and thirst.

The middle-aged woman tending bar ignores me. Only a few people are seated around the counter, and I watch them talk lightly with each other as they sip drinks and chew sandwiches. Every second spent sitting with no menu or water in front of me is like a slow off-kilter drumbeat. It burns my ears. If I could find a waitress in this town to call me "sweetie," smile, and take my order quickly, I'd tip her a full paycheck. I want kindness, goddammit, and I can't find it.

Another bartender appears from back of house. He's clean, well groomed, and handsome, about my age, maybe younger. He offers me a smile, displaying rows of even white teeth. "What can I get you?" he says.

I order a drink and ask for a menu. In a moment, they are both in front of me. "You must be a field hand," he says, not so much a question.

"Yeah," I tell him. "I move rigs. Where'd you come from?"

He smiles again. Easy, confident. "I'm from LA," he tells me.

"Just got here. I was doing cocktail service out there, so I'm doing this until I find an oil field job."

"What do you do?" I ask him.

"I'm a veteran. I signed up for a job services program that will give me CDL training for free. It starts in a couple months. Once I finish that I'll start looking."

"Don't wait," I tell him.

"What's that?" he says.

"Don't wait."

"No?"

"No."

He cocks his head to the side.

"The people that hire at these companies are so fucking stupid," I say. "They are so fucking dumb that they will have a stack of applications from qualified guys with CDLs and experience, and they'll just hire the first fucking guy that walks in on a day they happen to need somebody. The only way to get a job here is to knock on doors. They are all fucking idiots."

"Yeah," he says, "I've been watching YouTube videos of some guys who made the trek before and heard the same thing. I think I'll get better pay with a CDL."

"If you want to drive a truck," I say. "Otherwise, fuck it."

I hear the edge in my own voice, the hardness and directness. I feel my body draped over the bar stool with one leg hitched up and my elbow on the bar top now leaning over a plate of steak. I'm reminded of first getting into town, talking to an experienced hand who rattled off advice with a wolf's mouth. I was a tourist then, I'm a local now. It wasn't very long ago, but I feel that time. More and more these days I imagine staying here, settling down. It doesn't take long, I suppose, for a change to happen in a man.

THE WOODSHED

I'm getting sick of these motherfuckers." I'm on the phone with my brother Matthew. He's at his place in Los Angeles. I'm in my rented room, sitting on the bed and staring at the wall.

"I bet," he says.

"I'm in the break room the other day after work, filling out my time sheet," I tell him. "There's ten or twelve guys standing around, white guys, all telling racist jokes."

"Oh, man . . ."

"I tell 'em they're all a bunch of fuckin' peckerwoods and they should shut the fuck up, but you know, nobody pays any attention to me. I'm an 'angry garden gnome,' says Jack. You know how my beard grows in red? I've been wearing a winter cap . . ."

"You do look like a garden gnome," my brother admits.

"It's crazy making," I say. "There has got to be nothing more cowardly than a group of guys of one race standing around and talking shit about people of a different race. Nothing."

"That is some bullshit."

I'm fuming now. It has been bothering me for some time, building up inside me, a growing rage. "You should hear these fucking guys," I tell my brother. "A Native American dude I work with called Obama the N-word. I don't even know what to think about that. It never lets up. I have to hear that word like every other fucking day! And they're the patriots! These guys throw around this fake-ass pro-USA bullshit and then talk like fucking Nazis. It's making me nuts, man."

"That's frustrating," Matthew says. I'm stomping around my small room at this point, waving my arms around.

"I haven't been in a fistfight since I was fifteen years old," I tell Matthew, "but I swear to God, I'm ready to square my shoulders to some of these motherfuckers."

"Well," my brother says, "I'm not against boxing."

"I know you're not." I laugh. Matthew has studied Muay Thai for years. Now he teaches some classes. He's a true devotee. He got to it late and fought his first cage match at the age of forty.

"Taking somebody to the woodshed . . . ," he continues.

"Exactly," I say. "The woodshed."

"But only if, you know, it's a way you can settle something." Something in his voice stops me. It's calming. I quit pacing and flop on the bed.

In the past, I had always felt that Matthew started trouble for me, caused me problems. Growing up it was like he was my kid brother, and even deep into our twenties, I felt like it was always me cleaning up his mess, covering for him, saving his ass. In recent years, that dynamic had faded—we'd become friends—but my resentment lingered.

While I was in Williston, however, I knew I could express to him, more than anyone else, how much I was struggling. I just sensed it, that he could carry it, that I needed him to carry it. I was surprised at the careful way he listened and the measured way he doled out advice. This created a subtle sea change in our relationship. He was becoming not only more than my friend, but finally the older brother I wanted him to be, not just my rock, but an example to me of a good man, and the idea, perhaps, of a good hand.

"What do you mean?" I ask him. "You said, 'only if it's a way to settle something . . .' I'm talking about throwing hands."

"Some things you can settle like that," Matthew agrees. "'Let's step outside, get this done.' Other things, though, all you can do is walk away and shake your head. You gotta just figure God has it covered."

"I guess."

"That's what I think."

"Uh-huh."

"You aren't going to solve racism by punching a guy," he says. "And if you have that fight one time, well, you'd better be ready to have it every day."

COMPLICIT

The landlord opens the door with a wobble like he's drunk. It's the night before Halloween, and I half expect him to hand me a candy bar. He's got disheveled hair and an expansive smile. "C'mon in!" he says. In an amused way, I wonder if he's touched. I follow him inside. The house is a giant duplex with one set of apartments on the top floor and another set on the basement level. He leads me through the place.

"It's a good thing you came when you did," he says. "There are some other people on their way."

He shows me two spaces for rent, a tiny basement-level room with a single, small window for $795 a month and a slightly bigger room in the converted garage for $995 a month. The house has a kitchen and living room on both levels. I eye the kitchen hungrily. I haven't had a place to cook for months, and I'm starving.

Again, I'm hunting for a place to live. The boardinghouse is a step up from sleeping on the flophouse floor. But it doesn't have a kitchen. I am paying way too much money for food, and I am constantly eating garbage. I need a cheaper place where I can cook a vegetable.

The man introduces me to a couple in the downstairs living room, a woman in her late forties with blond, teased, hair-sprayed hair and her boyfriend, a mechanic in his midfifties. They smile and wave up at me from the couch.

"We really are like a family here," the landlord tells me wistfully.

"We're glad you don't look like a weirdo," the woman on the couch says. The man on the couch smiles blandly.

I tell the landlord I'll take the basement room, and I follow him to the upstairs kitchen. His arms jangle as we mount the stairs. Seated at the kitchen table, I'm looking over the lease and pulling out my checkbook when the doorbell rings. The landlord wobbles to the front door and opens it.

"Hello!" he says, his smile bright as ever. "Oh, hello. Come in, come in. I'm so sorry, though. But we just rented the room. You just missed it. I'm sorry."

A dark-skinned black man and woman have entered the house. Bundled against the cold and wind in fluffy coats and hats, they glance awkwardly around the living room. The landlord's smile is almost cultish, but he's scrambling, talking too fast. "I really am sorry. This gentleman here is signing the papers right now. See? You just missed it. That was the last room. I'm sorry."

The couple look at me, and I feel a spider of heat crawl up my neck and into my cheeks. My face flushes. I give the woman and man a nod in greeting, but I don't manage a smile. The woman wears a scarf of dazzling color.

"Oh, okay," she says. Still they stand there.

I know what I should do. I should tell the couple that a second room is available and that the man is lying. I should tell the landlord I don't want his stupid fuckin' room, and I should storm out. It is that simple.

But I've been looking for weeks. Again. Clicking on ads, driving to apartments, meeting weirdos in dirty group homes where the bathrooms smell like shit and the TV never turns off. But not this place. This place is good. It's clean. It has a kitchen. And I got here first. I found the ad, and I drove over immediately. That's how it is in Williston. "Shit is different up here," like Jesse once said. This is what you have to do to survive. I need a room and a kitchen as badly as anyone, black or white. I'm going to take it, goddammit.

"Terribly sorry," the landlord continues, "you've just missed it. So sorry you came all this way." He shuffles the couple out the door, and they are swallowed by the night. He shuts the door behind them and locks it, then joins me at the table, where he breathes a contented smile of relief.

"I guess you didn't want them in that second room," I look at him and say.

"Well," his smile takes on a sheepish quality, "I'm not really racist. But . . ."

I don't hear anything else he says. I stare at the check in front of me. Pale green. Rectangular. Written out to his name. This man isn't going to rent to that couple no matter what I say or do. I sign the check. He keeps talking. I sign the lease and hand it over with the check. I let myself out the back door.

I can say a lot of things about myself. About my moral sense, my ethical self, my belief in right and wrong. I can say I'm a good person. Maybe I am. But from that day to this day and forward, I have added a word to my sense of self, an addendum born in a life-defining instant. I am complicit.

GERMAN PRIDE

"Yeeeaaaah, I mean, I don't always tell the whole story." Jack Jackson sits across from me stirring his double Crown and Coke with a straw. "But you know that intersection in Florida?" he says.

"Where you got in all the fights," I say.

"There was a time down there, early nineties, that if I just got a bad look from a brother, I'd pull out my 9mm and unload it into their car, *bap-bap-bap*."

He takes in every inch of my face. He has told me before about the scuffles, the brawls, the fistfights, the knife fights, and the two times he's been shot. But this is the first time he's talked about using a gun himself, shooting into a car full of people? Maybe killing them? Because of a bad look?

A topless woman with a bored expression swings on a pole. It is late afternoon on a weekday, and Whispers is quiet. Jack and I have been at it for a few hours, bouncing from bar to bar, sucking on drinks, talking smack. We've been moving slow, though. Our joints ache, our hands ache, and our heads ache. It is an oddly introspective and quiet day.

A stripper walks up to the table. She's unsteady on her feet, has the rubbery look of a person abusing pills. "Either of you guys want a dance?"

"How much does it cost?" I ask, sipping my whiskey.

"Twenty dollars," she says, "or one fifty. Or two hundred and fifty, depending what you want."

Jack and I look at each other quizzically. "Maybe later," I say.

"What are those tattoos on your arms?" the woman asks Jack.

He's wearing a T-shirt, and he flexes his biceps for her. The word *Deutsch* curls around the thick muscle of his right arm, *Stolz* around the left.

"What's that mean?" she asks.

He looks at me with a smirk before he tells her. "German Pride," he says.

"Oh," she says, "you're German?"

Jack cocks his head to the side, "Noooooo."

"Oh," she says. She looks down, then back at Jack. "You want a dance?"

"No," he says, shaking his head. She starts walking away from us.

"Let me know if you change your mind," she says over her shoulder.

We watch her go. Then I see Jack exhale. His shoulders slump, and he stares for a long moment into his drink. It's a rare touch of silence for a man not much prone to introspection. He swings his head toward me, studies me for a moment. Then he arches his neck. "See this one?" he asks, pointing to a messy tattoo on his throat. It looks like a smeared pair of scissors.

"Yeah, what is it?" I swallow a mouthful of whiskey and Coke.

"Yeeeeeaaah," he says. He takes another sip of his drink. "Lightning bolts," he says. "SS."

I close my eyes for a moment. Then I open them. "I had them covered up," he tells me. I don't know what to say. I chew on a piece of ice and stare into my drink.

"This last stint I did, you know, the eighteen months in Washington State?" he says. "Well, my buddy, he's a Hilltop Crip, and I saw him just after I got transferred back there."

"A Hilltop Crip?"

"Yeah, a nigger. You know, in the Crips. The gang."

"Yeah, okay. I get it," I say.

"Yeeeaaah, we were friends, and I see him on my first day inside. I just got processed, and he's in the back of a transport truck. We wave at each other, 'Hey, how you doing, man! What's up?' Ya know. I don't give a fuck about a guy's paint job," Jack says, referring to the color of the man's skin. "He's, you know, we cut it up sometimes. We were friends." Jack shrugs.

"Uh-huh"

"Yeeeaaaaah." Jack whistles. "Well, the skinheads didn't take too kindly to that, oh boy."

"They saw you waving at him?"

"They didn't take too kindly to it, no, no, no." He sips his drink through the straw, then holds the glass in his lap with both hands, his head bowed like he's praying. "They paid a visit to my cell that night. Two of 'em. Big boys."

I study him until he looks up at me. He's half smiling. "You know who saved my ass?" he asks.

"Who?"

"The Muslims. Yeah, man, it was the Muslims! Skinheads show up to stomp me, and the fucking Muslims told them to lay off." He shakes his head, then pauses again. "Yeeeaaah, it worked. For a little while, anyway."

I take a big swallow of whiskey. Jack is a little drunk but steady. "I stayed unaffiliated for eighteen months of incarceration," he says. "Eighteen months. My son, who's locked up down south, even he said I was crazy for it."

"Jesus, man."

"Eighteen months. I'm sick of fighting," Jack says, ruefully shaking his head, then looking at me, then staring at his drink. "I'm sick of fighting."

I kind of believe him.

RAILHEAD IN WINTER

The sun meekly joins the sky and snow begins to fall. Ice collects in my beard. Bomani, the Congolese swamper I met over the summer, does the robot dance atop a railcar. His feet slide across the joints, and I take photographs. Since I last worked with him, Bomani has been joined in the rail yard by "Little Jim," a fellow Congolese.

Little Jim is well under five feet tall and barely speaks a lick of English. "Yes, yes," he'll say, and "Very good." He is paired off with a husky, good-natured white boy from Idaho called Big Jim.

"Most of the guys in the yard don't like working with him," Big Jim tells me of Little Jim, "'cause he's so small the only thing he can do is throw the boards. I don't mind though. I like having him there to throw the boards."

Little Jim has a big spirit. When he works on top of the cars, I hear him huffing and puffing as he picks up each board. He moves quickly and shouts every time he tosses one over.

"Him Big Jim," he says of his large, corn-fed friend. "Me Little Jim."

Afternoon comes on. The work slows down and the snow eases

up. A big-boned Native American swamper named Phillip, along with Big Jim, Bomani, and I, stand around as Little Jim tells us about Congo. His English is halting, and it takes him time to work out his ideas. Patiently, we help him along. It is a rare gathering of hands that doesn't include a yelling alpha male. Little Jim tells us about the war. "Very bad," he says, "very bad."

His language may be limited, but his face tells a tale. It is a youthful face ravaged by grief. He lived in refugee camps, his gloved fingers show us, for nine years. When he gained asylum in the United States, he settled first in Idaho before moving to Williston. His wife and children remain in the camps, he tells us. He is working to bring them to the United States.

We watch a video on Little Jim's phone that tells us about the conflict in Congo. Bomani watches quietly. He then tells us that when he lived in Atlanta, he watched *Roots* on TV. The next day he went to work, and his boss called him a nigger.

"If I had not seen Kunta Kinte on TV the night before, I would have let it go," he tells us. "But I saw Kunta Kinte whipped." He recites a line from the movie, "'My name is Kunta Kinte!'" Bomani shouts, "'My name is Kunta Kinte!' So, I sued the company for a great deal of money." We exchange a good laugh in approval. "I did. I made very much money." Bomani nods, "But here. Here I think it is okay if some people call me nigger."

Phillip doesn't like the sound of that. "I don't think so," he says.

"No," says Bomani. "They say it, and they don't know what it means. They are dumb. They think I'm a nigger? I don't care. One time an Indian in the pipe yard called me nigger, and I got very angry. But I realized that was okay because I am brother with all Indians."

The Native American swamper lets out a hearty laugh.

A VOICE MAIL

Hello? Hello? Eh, this is, uh, Sonny Smith. I'm calling for, well, I hope this is Michael, I can't tell from the damn answering message. But uh, Michael, this is your father. I just wanted you to know, there's a guy down the street here from me. He needs his basement cleaned out. I don't know what your work situation is right now, but uh, it seems like this guy could use a hand and, well, there might be some money in it for you if you wanted to help him. He's doing some construction things, renovation things, too. I think. Anyway, I just thought that, uh, you might need the work. So, if you want to make some bucks, gimme a call. Also, uh, happy Thanksgiving, that was last week, I guess. Hopefully you get this message. Hell, I don't know. Anyhow, this is Sonny Smith, and my number is . . ."

BROTHERS

I drop by Huck's apartment one morning, and as he takes apart and cleans his handgun, I grab his acoustic guitar and play a couple songs. He looks at me differently when I play, and I can see something shift in his understanding of me. "You are a very determined person," Huck says slowly, slightly emphasizing each word. I never had any natural talent as a guitar player. I argued with the instrument for hours every day for years only to see friends pick it up and nonchalantly surpass my skill level within a matter of months. But sometimes turtle beats hare. My abilities as a rhythm guitarist—and some may still describe them as modest—are wholly earned, and a source of great personal pride. Huck's simple statement is one of the most insightful things anyone has ever said about me.

When he finishes cleaning his .45, I put the guitar back in its case. We toss his guns—a shotgun and the handgun—into the back of the Blazer and drive out to his parents' trailer, about twenty minutes west of Williston.

We get to his folks' trailer, set by itself down a windy road surrounded by switchgrass, and go about setting up targets. We find an old pumpkin, and a safety orange-colored hard hat I had sitting in the back of my Chevy. Huck hands me the .45. At his apartment, I had noticed how fastidious and thoughtful he was in the way he handled the gun. Holding the .45 in my hand, I take a moment to register its weight. The gun feels like responsibility, and it feels like power.

My first attempt is a kill shot. At about 20 yards, I blast a hole directly through the top of the hard hat. I empty the rest of the clip, but my beginner's luck doesn't hold. We take turns passing the pistol and the shotgun back and forth, blasting away at the pumpkin and the hard hat. The bang of the weapons is a thrill. The smell of gunpowder lingers in the cold air. Ammunition is pretty expensive, though, and we finish quickly. I collect a shotgun casing from the frozen ground and, for some reason, stick it in my pocket.

Huck's mom comes home from the store, and we help her in with her groceries. The trailer is small but homey. Inside of it, Huck is deferential to his mother, but he likes to cuss in front of her, I can tell, because it gets on her nerves a little bit. I look at a photo on the fridge of Huck as a boy posing with his parents. His mom smiles while his father stands stoically gazing forward. Huck looks bewildered. After some small talk, we say bye to his mom and head back into town for beer and wings.

It is usually difficult to pinpoint the moment when a casual friendship develops into something deeper, but for Huck and me, it was on this day. After shooting, we find ourselves tossing back pints of amber, our fingers and mouths thick with Buffalo sauce, and we began goofing off each other in a way that I haven't experienced with anyone since grade school.

Huck was growing more serious as his legal troubles mounted. He'd begun confiding in me. I was older and worldly, and I didn't write him off as an idiot like most of the guys he knew. It wasn't that I didn't find some of his antics stupid; it was just that I felt attuned to Huck's incredible sensitivity. He was like me in that way, tuned in to the emotional space around him at a very high frequency. He felt deeply and had trouble pretending otherwise.

Huck's innocence had certainly taken a beating over the past several months, in parking lots, police cruisers, and jail cells, but he hadn't by any means lost his shine. The guy was a five-year-old every time he took a lick of ice cream, his smile a tractor beam.

No one has ever got me laughing like Huck before or since. We laughed until our ribs were cracking, laughed like a river pouring through us, laughed so hard as to clean ourselves with laughter. We sat at the bar, roared, and shook with it.

Huck, wiping the sauce from his chin, got the attention of the bartender. "You know, we're brothers," he told her proudly.

"Oh yeah?" the bartender said, looking back and forth between us. "I can see that."

Sixteen years apart in age, we couldn't look more different. Huck is a full twelve inches taller and seventy pounds heavier than me. My hair straight, his hair curly. My eyes blue, his eyes brown. His face full and fleshy, my nose sharp and cheekbones prominent. And yet. And yet the words as they rolled off Huck's lips were much more than a joke at the expense of an unknowing stranger. He meant it, and if the bartender felt any hesitation or any reason to disbelieve him, I could not sense it.

There was something between Huck and me. Something men, I think, have trouble talking about. We weren't related. But somehow we were blood, the same blood I share with my family. But whereas that blood was given at birth, with Huck that blood grew between us. It was a kernel embedded in our skin that first hot day I got dust in my mouth on Sidewinder Canebrake 103, and it blossomed into a flower in our shared vein. Once it was there, I felt that blood every bit as strongly as I'd feel it walking into a family reunion. Huck and I were brothers.

So, like pairs of brothers all across the globe, we decide to do some dumb shit together. We head to the strip club.

ROSALIE

At the front door of Heartbreakers, the bouncer asks Huck and me if we have any weapons on us. I think of the guns in the Blazer, then I tell him I might have a pocketknife. He pats me down. I'd left the knife at home, and he lets us through.

We are immediately set upon by two dancers. A tall black woman pulls Huck into a curtained booth and offers him a blow job. I'm accosted by an Asian woman, probably in her late thirties, very skinny with big fake tits. For twenty dollars, behind a dingy red curtain, she twists on top of me like greed. I pay her money to leave me alone. "No more dances, but thanks," I say, escaping her for the bar.

I suck on my beer with my back to the room. I've already had a good bit to drink. I can feel the alcohol working through my bloodstream and into my brain as Huck ping-pongs around the club talking to the dancers. None of them are too pretty, but it is early and the club is only now starting to fill up. I take another swallow and turn back toward the room. A raven-haired woman in a bikini struts onto the stage as the opening chords of "Pour Some Sugar on Me" reverberate out of the speakers. The dancer grabs the pole and swings just as the guitar kicks in, her long dark hair tracing a wide arc under undulating red lights.

My first serious girlfriend worked as a stripper. We met my first semester at college. I was a boy at eighteen and a desperately late bloomer, or so I thought. She was a worldly nineteen-year-old. She asked me on a date, and we sat on a dock in Baltimore's harbor where I read my poetry to her out loud. I lied about my sexual experience—I had none, but she taught me—and we fell in love like stupid, crazy kids. She quit dancing and went on to get her master's degree in women's studies. Now she's married, with kids, and runs an arts organization in Topeka. Occasionally, we still call each other up and have a good laugh.

Several of her friends worked the clubs. Some of them did "bottle service," meaning they slept with men for money, and some of them didn't. Some of them performed outcall service, meaning they fucked guys in hotel rooms, and some of them didn't. Some of them had suffered sexual abuse as children, and some of them had grown up in loving homes with supportive parents. Some of them took pills; most of them didn't. They were like the other young women I knew at that age except they were earthier, had sharper senses of humor, and they were openly horny. To a repressed, farm-raised Catholic kid whose mother didn't allow him to take sex-ed in high school, these women were a revelation. I don't exaggerate when I say they changed my life.

Since then, I've had several friendships with and even dated a few sex workers. There is darkness in that world, of course, but I'm no stranger to darkness. It has never been a deal breaker to me if a friend of mine, or a lover, at some point traded sex—or the idea of sex—for money.

On the stage, the raven-haired dancer is upside down on the pole with her legs spread eagle. She slides down toward the dollar bills on the floor as Def Leppard rides the chorus into the fade-out. I down half my beer in a swallow, then set it down.

A tall woman appears in front of me wrapped in a winter coat. Her bright eyes catch mine, and she reaches her arm out dramatically. Just as a new song begins to play, I take her hand in a display of mock gentility, and she twirls into my arms. I hold her there for a moment and then swing her out, and we dance.

In the center of the room, in the middle of a thick group of work-uniform-clad horndogs in ball caps and the soft, enticing uncovered flesh of women, we dance. Her winter jacket is bulky, and her hood is up, hiding her hair from sight. I lead for a few steps. She is very pretty, her face perfectly symmetrical. She holds my eyes and says, of my dancing, "You know what you're doing." I grin. I barely know what I'm doing, but yeah, sure, I'll take it.

"What's your name?" I ask.

"Kelly," she says, but then quickly, "I mean Rosalie." My face becomes a question mark. She lets go of my hand and disappears into the back of the club.

I watch her go. Then I turn my back on the men around me, the chumps, the losers, the watchers. I don't want to meet their eyes, see their expressions. I don't want the taint of their opinions to scuff up my moment. I'm wearing a shit-eating grin when I return to the cool comfort of my beer.

A few minutes later, a woman appears next to me with long straight brown hair wearing a string bikini bottom and a torn black rock-and-roll T-shirt that exposes a pair of perfect breasts. She is tall in stiletto heels and wears a wonderful smile.

"What's your name?" I ask impulsively.

"Rosalie," she says before I realize it is the woman with whom I just danced.

"Where are you from?" I ask.

"Who cares?" she says. I nod in agreement. Who gives a shit? Her lips are red and her legs are miles long. She'd be a knockout in any room. She grips my forearm. "There's a poet called Charles Bukowski who I love," she tells me, and she leans into me to whisper his "Love Poem to a Stripper," her lips just grazing my ear.

> 50 years ago I watched the girls
> shake it and strip
> at The Burbank and The Follies
> and it was very sad
> and very dramatic . . .

She recites the words carefully, deliberately, and her wet breath warms my ear and neck. I close my eyes to listen and to breathe her in.

> Rosalie was the
> best, she knew how,
> and we twisted in our seats and
> made sounds
> as Rosalie brought magic
> to the lonely . . .

I feel alone with her in this crowded room, and the intimacy is bracing. This has to be one of the weirdest and most erotic things that has ever happened to me. Bukowski, the poet laureate of strippers and drunks, seems to be blessing us.

She recites the entirety of the poem. When she finishes, Rosalie turns away from me without another word and strolls off. All eyes are on her. She is young and stoned with a careless beauty that a man could kill a marriage for. Some probably have. I watch her cross the room, one long, practiced step after another, picking up and tossing off a thousand desires from end to end.

Huck watches her, too. He's like a Looney Tunes character with his eyes bulging and his tongue on the floor. I sit alone on the stool and curse my drunken brain for not having a poem to recite back to her. Why don't I know any fucking poems? I try to think up some, I don't know, Robert Frost . . . *two roads in a yellow wood* . . . is that how it goes? And I watch the men pour money down Rosalie as she works the room. Already, this woman and the wonderful, intimate moment we just shared, only a few moments ago, are a million miles away.

Another woman appears, a hot showgirl from Vegas with a long blond wig and a wicked smile. The lines beneath her makeup attract me to her. She is older than Rosalie, and devilish. She takes my hand and leads me to the back room. Her smell is perfume and sex. She crawls on top of me, and her taut body feels full to bursting. She puts her hands under my shirt. "Ooooh," she says, "long underwear," and it provokes in me such a good, full-hearted laugh that she joins in. Her laughter is sensual and hoarse with no meanness in it. And she moves her body on top of me without a trace of a hustle, and I melt under her thick thighs and slender fingers. The song ends, and I give her more money. The next song ends, and I give her more money. The song after that ends and, flushed and laughing, we wrap our arms around each other in a hug, and I thank her.

When we leave the curtained booth my heart is slamming against my chest. I've fallen in love twice in one night.

I rejoin Huck at the bar. At this point, we are both shit canned. We stagger and sway and slur at each other, red-faced and goofing. Rosalie takes the stage to dance, and we sit directly in front of her, sliding dollars into her G-string. She writhes in front of us, her body bending like a slender pine in a hard wind. She climbs the pole above us, swings on it, then plants her feet and drops to her knees in front of Huck. He slides a dollar up her thigh and closer and closer still, and even closer until she slaps his hand away, swings up to a standing position, and puts a finger in his face. "Don't fucking touch me there."

A bouncer puts a hand on Huck's shoulder. "You tip well," the bouncer says, "so I don't want to throw you out."

Huck looks at me bewildered. "I didn't mean to . . . ," he says.

"Yeah, you did," I say.

Rosalie crawls across the stage toward another group of customers. The spell is broken. The night is done. "You're a fuckin' idiot," I tell Huck. He protests his innocence, but I don't believe him. "Forget it," I say. I'm not sure how much I care, how much I should care, or how much caring I've got left. I could be angry, or I could accept that this is just who Huck is, and this is who I am. We are both wrong. We are probably always wrong all the time. But I don't want to dig into the mechanics of my heart right now, I don't want to ask myself any hard questions tonight. The evening ends like a bite of a sour apple.

JACK COMES BACK

My feet crunch in the snow as I walk to the car, my hand wrapped around a can of beer, shoulders hunched in the cold, shimmering puffs of breath appearing and disappearing in front of me. The sky is a pasty gray cloud, and the temperature hovers just above zero. It is late December in North Dakota.

I open the rear passenger-side door and slide into the car next to Jack. He's yammering on about the two women he's been banging since he got arrested in Washington State. A tallboy of Coors Light sits between his legs.

"Yeeaah," says Jack, "this one chick I'm fucking, she's a freak, man. Yeeaah, she keeps texting me, man, sending me photos of her pussy. Yeah. This other chick . . ."

Huck sits in the passenger seat in front of me, his legs pushed against the dash, his knees up near his head.

The guy driving has his seat slung so far back he is almost lying down. In his midtwenties, with a flat-brimmed designer baseball cap sideways on his head, he chews compulsively on a silver stud below his lip. I've never seen him before. We pull out of the parking space and move down the alley and through the subdivision.

When Jack finishes his monologue, he turns to me, "How you doing, bud?"

I am doing barely okay. The past few weeks have been rough. The cold and the dark are taking a toll on my psyche, each day grinding me down a little more than the last. Two weeks ago, after our first dusting of snow, the brakes on my Chevy locked up when I hit a sheet of ice, which sent me hurtling through a stop sign and into oncoming traffic. I barely avoided a head-on collision and ended up slamming into a parked Oldsmobile. This being Williston, the local mechanics are all booked up; no one can look at the Blazer for six weeks. So, I've got no wheels. On top of that, the job has gotten more dangerous in the snow—everything is slippery, visibility is bad; I find the fear I felt those first weeks in the patch rekindling itself in my belly. A few days earlier, a ladder slid out from under me as I connected some tubing between mud equipment, fifteen feet off the frozen ground. I clung to a steel beam, my legs flailing beneath me until a floor hand sauntered over and repositioned the ladder. He strolled off before I could say thanks.

"I am doing alright," I tell Jack. It is good to see his busted-up smile and sharp eyes. Broken and ragged as he is, he's buoyant. Maybe he can't win, but he can't be beat either.

Jack had headed back to Washington State to see family, and he was pulled over just outside of his hometown. The police forced him from the car at gunpoint, and he was thrown in prison for violating probation. After a week in lockup, he was released, but the damage had been done: Diamondback fired him, and Jack was back in The System. Now he was smuggling weed into Williston. He wasn't legally allowed to leave Washington State but he had no other means of income. He felt that selling pot was his only chance to put some money in his pocket.

We drive downtown. The shops are strung with Christmas lights. Paper snowflakes and plastic Santas fill the windows. The driver stops the car at the corner of Broadway and Main, and Jack hands him a Ziploc baggie from his coat pocket. As the driver gets out of the car to make a deal, Huck leans his seat all the way back so his head is all but resting in my lap. He looks up at me, "How ya doing, bud?" When the driver gets back in the car, we get moving again. Maybe it is the weather, but the mood is subdued. I look out the windows at the Christmas decorations—they appear festive in the evening, but in daylight everything seems washed-out, kind of dingy. The car drifts through gray streets beneath a gray sky.

"How much money you think I could get for this car?" the driver asks Huck.

"I dunno," Huck says, "maybe thirty-five hundred?"

The driver chews on the stud below his lip for a long moment. "I'd give it to you for twenty-five," he tells Huck.

"Oh. Well, I got a truck, bud," Huck says, surprised. "But I can ask around for you, I guess." The driver nods and continues chewing on his chin stud.

"I've got a Mustang in Florida," he tells us, "but I can't get to it 'cause they got a warrant out for me. I can't even go back to any of the surrounding states really. Might get locked up."

"Yeeeaaah," says Jack, exhaling in agreement.

BUFFALO WILD WINGS is loud and full of TVs. The driver doesn't order any food, and he barely touches his drink. Almost immediately upon walking inside, he walks outside to talk on the phone.

"Who is this guy?" I ask Jack.

"Friends with Slim," Jack tells me. "You know, the parts manager at Diamondback."

"Don't think I know him," I say. We pick at our wings and sip our beers, half hanging out and half staring at the TVs that line the walls. When the driver comes back in, he asks Huck again about buying his car.

"You could just use it for parts," he tells Huck, "for two thousand even."

"I don't think I need those parts, bud," Huck tells him.

I'm on days off. I've got a whole afternoon to drink away. I can't think of anything better to do. But Jack needs to swing back by Slim's, so we leave the bar to a sparsely falling snow. The car winds its way through town, new snow settling on top of the same old snow, sticking to the roads, already covered by a thin sheen of ice.

As we approach Slim's house, the driver sees an open parking space on our left, directly in front of the apartment. He maintains speed until we are right beside the space, and then he yanks up on the emergency brake. The car swivels in a tight arc on the ice, cars and trees swirling past. Blood and adrenaline rushing to my brain, time slows to lazy, and all our thoughts and chatter shrink to the head of a needle. We spin a full 180 degrees and slide into the empty space, tucked perfectly between two parked cars. All at once, the bubble bursts into laughter. "That's some good fucking driving!" Everybody talks at once, looking at each other and laughing. The driver proudly grins. We pile out of the car.

SLIM LIVES IN a split-level townhouse with his wife and two kids and another couple and their kids. Jack enters talking and keeps on talking. He moves up the stairs with his big shoulders hunched, his bright eyes taking everything in. A gaggle of children cavort in the kitchen with one of Slim's housemates, a small woman—a girl, really, in her early twenties—with bad skin and stringy blond hair. The kids dance and play in a loud, wild jumble at her feet. Jack flirts with the mama.

"Why don't you come out and get a drink with us," he says.

"I'm watching the kids," she tells him.

"What kids?" Jack asks innocently. They race around him.

I stand uneasily in the living room with Huck and the driver. We're out of place here. I don't know what to do with the children, how to interact with them. How many are there? Four? Five? I literally start to sweat.

Jack goofs with the kids. He calls a little boy a "sissy la-la" for crying and chases a little girl with a big stuffed bear. Then he sidles up to mama as she wipes jam off a baby's face. He leans close into her, his hand moving unconsciously up to his mouth to cover his busted teeth, and says something no one else can hear. She glances sideways at him and lets loose a fantastic smile before shaking her head and tending again to the baby. Jack leaves her in the kitchen to mug for us in the living room. He's acting like the place is his, like the woman is his, and like the kids are somebody else's.

I take a seat on the couch. Huck and the driver hunker down next to me. *Sponge Bob* is on the set.

"Would you want to buy these sunglasses?" the driver asks Huck. "They're designer, and the lenses flip up, see." He shows Huck how the lenses flip up.

"Um, they might be good for a crane operator," Huck tells him. "Maybe Charley would want them."

The driver looks at me and then back at Huck. "Charley?"

"We work with him."

"You wanna call and see if he wants the sunglasses?"

I take another look at the driver. He appears pretty collected, sitting on the couch in his baseball cap. In fact, he's incredibly focused. I watch him gnaw the piercing beneath his lip, and I suddenly feel foolish for missing it earlier—this guy isn't trying to sell off everything he owns for no reason.

Slim arrives home from work. He is tall and lean as his name suggests, with neck tattoos reaching up toward a pronounced Adam's apple. He wears a black leather jacket and a black baseball cap over dyed black hair and a pale, serious face. When he enters the living room, he exchanges a few hushed words with the driver, and the driver slips out the back door.

The kids are shuttled downstairs by their mama, who disappears after them. Jack and Slim immediately unload several pounds of weed onto the kitchen counter and begin dividing it up and weighing it out. How the fuck did I end up here? I thought we were on our way to a bar.

Outside, night falls as Huck and I sit quietly on the sofa waiting for Jack and Slim to finish their business. I check the temperature on my phone. It has dropped to negative 20 degrees. My apartment is only about a mile down the road, but I'm not wearing any of my arctic gear and the distance feels insurmountable. The walk would be brutal, and if I slipped on the ice walking home and cracked my head, I'd freeze to death. But I want to leave.

"How 'bout we grab a beer?" Jack finally says, as if I hadn't been begging for hours.

"Let's fuckin' go," I say.

We pull our hoods over our skullcaps, zip our coats to our

collars, and Jack, Huck, and I walk out the back door and down the steps behind Slim's townhouse. It is dark, and the pancake-thick layer of snow on the ground has an added crunch to it.

In the alley behind the house sits Huck's '68 powder-blue Chevy pickup. The three of us pile onto the bench seat with me in the middle, windows fogging over as we stamp our feet and blow into our hands. There's no working heater. Huck cranks the engine to a start (there's a trick to it) and we get moving. Air spills through the holes in the rusty frame. I feel it crawl up my leg like liquid nitrogen. It is the kind of cold that clenches your whole body over itself like a fist. We're soon giddy at the sheer audacity of the temperature, laughing and hollering and slapping our hands, silver coming out of our mouths in glittering, glistening puffs. You can see more clearly in that kind of cold, the air so thin that every image pops against the black wall of night. Things suddenly feel right. I'm happy to be out of Slim's house and with my two friends. We jabber as the truck rattles along to the bar, through streets now twinkling brightly with festive lights of Christmas strung between the houses, plastic reindeer, and manger scenes glowing warmly in front yards.

AT THE BAR, Jack pulls a wad of cash the size of a gorilla fist out of his jacket pocket, a Ben Franklin wrapped ostentatiously around the fold. He buys us all a round of double whiskey and Cokes, and we settle into a well-worn groove of friendly banter, the ice melting off our coats, fat drops splashing onto the barroom floor. We joke about work, money, and women. Then Jack starts in on something, and I can't figure out what it is he is talking about.

"I didn't wanna do it," he says. "Yeaaaahhh, it's not something I'd usually do. I don't mess with that shit. But my cousin could get it. Yeeeaaah, I don't like it, but, you know, as a favor, I told Slim I could do it one time. Just one time."

In addition to the weed, Jack explains, he brought half an ounce of heroin with him from Washington State for Slim and Slim's wife.

Suddenly, I feel profoundly tired, but before I have time to process this information—to dwell on the children in the house—Jack is veering into a monologue about his conquests in Washington, the two women he's fucking—neighbors or cousins or sisters or whatever they are. "Yeeeaaaah," he says, "I've got one of them," and he puts his thumb and index finger about an inch apart, "this close to letting me sell her ass."

I nearly spit my drink out laughing. I feel my back loosen and my body relax. Jack motherfucking Jackson, always saying some funny crazy shit.

"Yeeahh," he says, encouraged by my laughter, "I'm serious, man. She'll be a cash cow. Yeeeaahh, I'll show you." He starts messing with his phone. "Lemme show you this video of her sucking my dick."

"I don't wanna see that," I say.

"Yeeaahh, man. She's got skills," he says. He looks at Huck and nods, "She'd be good for you." He looks at me. "She'd be good for Huck," he says. He says he thinks her skills are worth $100 a pop. "I keep fucking with her," he tells us, "acting real sweet, telling her I love her sometimes and then calling her a filthy fuckin' whore. I'm close to breaking her." He nods, pleased with himself.

The worm that ruled my brain during those first months as a field hand slithers through, down deep into the soil of my belly. I feel it grow and coil through my intestines and into my stomach, long, heavy, and putrid. I slurp on my drink. Jack can't get the dick-sucking videos to load. He gives up.

. . .

WE LEAVE THE bar drunk, stumble across the white crunch of the parking lot, and relive the ritual of Huck's freezing jalopy. I take photos with my phone of Jack looking sinister in a black hood. Shadows thrown across his face, he scowls into the camera. I take photos of Huck putting on his cool face, looking stoned while smoking cigarettes.

Within a year Slim will lose his job, and he and his wife will be found dead of a heroin overdose. I'll never find out what happened to all those kids.

THE COLD

Big, fat white flakes slant through the diffuse light of day, cascading, tumbling one over the other, silently, gently alighting onto the frozen ground and sweating off the idling machines. Christmas has come and gone. After a brief trip to Maryland for the holidays, I returned to Williston in early January in the midst of a brutal cold snap—negative 50 degrees with the wind chill. My trip back east had softened me. Life was simply easier there. The weather was more temperate, my friends were gentler, and I was surrounded by family. Boarding the plane back to Williston, I struggled to get my mind right. When I landed, I turned off my phone and took to the warm bed of my rented room, eating frozen pizzas and streaming movies on my laptop for several days, like a sad, fat pajama-wearing zombie. But when I check my bank account, I despair—work was slow before Christmas. I call dispatch and they send me back into the field.

Every driver in the region attaches a heater to his vehicle's engine block, plugging cars and trucks in at night so pistons will spark in the morning. At gas stations and truck stops no one locks their cars anymore. Drivers head inside for coffee and snacks as silver plumes of exhaust fall up out of mufflers, sparkling in the frigid air. It takes a little getting used to, not powering down the crane pickup when we pile out for breakfast. But nothing much affects nothing much. We get to location, and on location, through blazing sun or falling snow, we work.

Most guys wear specialized gear: long underwear under puffy arctic weather-resistant boots, lined winter jumpsuits, and gloves. Hard hats sit on top of hoods or caps. For winter face masks, almost all the men have chosen the same evil-looking skeleton design. It's a roughneck Día de los Muertos.

I stand in the snow with a couple jolly skull-faced roughnecks, and we watch a new crane operator, a worm, attempt what should be a simple move. But he swings the boom wildly left and sends a group of workers running for their lives. The worm started after Yeah Buddy was run off, and he's been terrifying swampers and operators ever since. I find it galling that he hasn't been fired.

"Too bad he didn't die on, I dunno, Christmas," I tell the jolly roughnecks. They agree wholeheartedly. The thought kind of cheers everybody up.

"I'd do something else," one of them says, "but I've only got a seventh-grade education. I left school to do tree work, man."

"That sucks," I say, and again they happily agree. I'm bummed when I get called away to do something. "See you later, I guess," I say, stomping off through the tundra.

LATER, ON MY lunch break, I eat a ham and cheese sandwich alone in the pickup and stare out the window at a group of hands wrestling in the snow. I imagine they are the jolly pair I'd been talking to earlier, but it is impossible to tell. They grapple each other to the flat, frozen earth with fat, juicy snowflakes flailing right along beside them.

OVER THE WEEKS that follow, the temperature rotates around a core of 0 degrees, swinging on a narrow pendulum from about 18 above to 18 below. I check the weather every morning before work, and I'm astonished to find myself exuberant when the mercury reaches into the positive teens. As long as I'm moving, my body core at that temperature will warm up. In the negative teens and colder, that is not the case. No matter how fast you move or how many layers of protective gear you wear, at negative 18 or negative 20 degrees the cold will get in your bones and stay there.

Roughnecks carry torches and we spend hours pushing back the encroaching ice with flames as it threatens to swallow whole sections of steel. As simple a job as loading up mats becomes a fraught and exhausting exercise as I bang at the mat's holes, using a shit hook to dig the frost out before sliding it in and signaling the crane.

Some days are sunny, and despite the freezing temperatures the ice melts into rivulets of frigid water, cold and wet as a mountain stream. When the water pools, it becomes impossible to tell the liquid from the ice. The glint of the sun off the glaze plays tricks on the eye, and a wrong step leaves me splashing through it.

I find myself bashing at the corners of mud tanks, generators, and mats, trying to free the negative spaces—sockets, eyeholes, and joints—of ice while sloshing through frozen water, suddenly up to the ankles, threatening my legs and feet with frostbite. Frostbite is a common problem in this weather. The arctic boots do an amazing job, but there is no room for error. Simply sitting too long in the truck, allowing your feet to sweat, can leave a man walking around with frozen water between his toes. And while everyone does their best to cover any exposed skin, I'll meet several men who are missing the small, fleshy part of their nose, just after the cartilage ends but just before the tip, as if winter's skinny fingers had pinched them there.

The trucks and cranes now run 24-7, churning through diesel overnight, the danger they won't start in the morning superseding any desire for conservation. Still, the cranes begin to break down regularly. The trucks sometimes refuse to start; but nothing stops. Progress, irrevocable progress, moves forward like a marching army.

My Chevy is declared totaled. I rent a car and head to Minot to buy a new vehicle. On the drive, I see the results of five different car accidents, including a flipped RV and a jackknifed trailer. They lie crumpled on the shoulder of the highway like slain enemies on the road to the castle of the Winter King. In Minot, as if in a country song, I say goodbye to my Chevy and hello to a Ford, a 2007 gunmetal gray 4-cylinder Ranger pickup. She and I will eventually see almost all of the country together.

Not long after I buy her, I drive to the yard one morning in the throes of a full-on whiteout. The road in front of me appears and disappears, waves of white washing across my field of vision. Snow rolls and tumbles, pushed in gusts from every direction. Sheets of white swing one after the other across my eyes. I can see sometimes no farther than a few feet in front of my truck. Then suddenly twenty or thirty feet ahead, a car will appear, blinkers on, inching forward, slowly, carefully tracing the lines on the highway. Then it disappears. Time gets funny in this cocoon of white. I move through it, a disintegration of sorts, a disappearing, a breaking down. White turns to white turns to white again.

When I arrive at the yard, dispatch is all but empty except for

Bobby Lee, Huck, a dispatcher, and a young hand we sometimes call Smurf. The group of us hunker down in the wobbly office chairs and warm ourselves with comradery, sly chatter, field jargon, and the shared language of bad jokes. Bobby holds court with his legs extended and his boots crossed. Huck this morning is at his aw-shucks best—that peculiar strain of innocence shining in his eyes as he hangs on Bobby's words; he's not trying to impress anybody, he's simply listening. The dispatcher is looser than I've ever seen him. His lip fat with dip, he makes a crack about the condition of Smurf's gloves before digging up a new pair. Even though there will be no work today, I'm glad I woke up and made the drive. We slouch around for thirty minutes or so, the snow dripping off our jackets and hats and wetting the floor, talking most intently about nothing. Then we head back into the beauty.

THE SIDE OF THE ROAD

Unlike the other operators, the Gruff Crane Operator drives his swampers around. "You're standing in the cold all fucking day. I figure the least I can do is drive the fucking truck," he barks. We are about to leave the yard, and there is an issue with power running to the trailer attached to the pickup. He gets out to fiddle with the cables. A fat bed truck driver sits in the passenger seat, and I'm stretched out in the back, a rare luxury. I haven't pulled my winter gear over my clothes yet, so I let the Gruff Crane Operator work on the trailer by himself, but I feel a twinge of embarrassment: a good hand would be helping. Before I grow too guilty, he slides back into the truck and we start moving.

A few miles down the road, the crane operator notices the trailer is still having problems. He pulls off onto the shoulder and hops out. I hesitate a moment, but resigned to my duties, I zip up my jacket and leave the truck. I walk down the passenger side of the vehicle and around the back of the trailer. The Gruff Crane Operator isn't there. I'm confused for only a moment when the brake lights turn off, and the truck starts moving without me. "Hey, wait up!" I holler, chasing after it. The truck pulls onto the highway and heads off down the road. I'm waving and jumping and whooping as the taillights fade down the stretch of country highway.

It is 20 degrees below zero, and the sun has yet to rise. My breath crystallizes in front of me. I watch the crystals all but fall to the earth as soon as they appear. I look up. The sky is racked with stars. They pulse in brilliance through an atmosphere so thin and clean that I feel like if I reached up to grab them, they would burn my hand. But they wouldn't—they are cold and distant—so I don't. I stuff my hands in my jacket pockets as a single big rig barrels past me. It takes all the sound with it, and the silence once it passes is full and strange. I start walking, my boots on icy gravel the only sound.

The moments are fat, and I study them as I would a precious stone, turning them over in my mind's eye both for their exquisite beauty and in investigation of their flaws. Under stars in this desolate place of such great silence, I feel minuscule and lonesome. Here I walk on a small, blue planet in a galaxy that rolls onward and onward past the eternity of prairie that surrounds me.

With every second the cold grows outside and inside. It seeps up above my socks and claws at my ankles. It burrows through the zipper of my coat and snakes its way down the tender flesh of my neck. My feet rapidly begin to lose all feeling. The town of Ray is probably twenty miles' distance, and the crunch of each step marks a meter toward that unobtainable goal. I'd never make it that far in this weather.

I'm as awake as I've ever been, every second a gift unwrapping. Someone will pick me up, I know that. But for a moment, I'm as wild and free as a coyote, kicking down the dark highway before daybreak, my heart banging out a tattoo on my chest, death licking at my heels, the sky so big and open I could fall right through it.

The truck pulls ahead in front of me, and I hop in.

"Where the fuck you been?" asks the Gruff Crane Operator. "Decided to take a fucking walk, eh?"

CHARLEY, HUCK, AND ME

I hate the term 'everything happens for a reason,'" Charley says. "People just say that because of tragedy. But no, everything doesn't happen for a reason. Things just happen. And then you're fucked. Is there a reason you're fucked? No. There's not. Maybe you zigged when you should have zagged, but trust me, there's no *reason* beyond that. No higher power gives a fuck."

Charley is a crane operator with a fatalistic streak. He's thick, with close-cropped hair, a trimmed beard, and, unlike any other field hand I've met, designer clothes. At thirty-three years old, he's the same age as the crucified Christ, and he likes to point that out. Charley himself died and was reborn one day when a pipe rack fell on him in Diamondback's yard.

"It landed across my back, right above my fucking pelvis," Charley tells Huck and me. "It knocked my wind out. Pushed out everything that might have been inside of me. I excreted every drop and every ounce. Do you hear what I'm saying? It made me shit my pants! My guts were literally hanging out of my asshole."

It is early morning in mid-January—the sun has yet to rise—and we are driving to a work site east of Tioga. Leaving Williston, we pass dozens of gas flares. Arctic conditions have created a strange change in the light coming from the flares. The air is packed with tiny ice crystals, and the surfaces of the crystals act like a thousand small mirrors. Instead of illuminating a circle around a flare like fire does in normal temperatures, the ice crystals shoot light directly upward in an extended vertical pillar. It is a haunting image, these luminescent bands stretching up into the dark sky. We pass towers of light on either side of us; halos. Some disappear behind while others keep popping up on the road ahead.

"I tried to pull myself up," Charley continues. "Nothing. One of my lungs had collapsed. I tried yelling help. Nothing came out. I knew I was fucked. So, I laid my face down in the dirt. How am I going to get out of this?

"People ask me, 'Did you pray?' Ha! Did I *pray*? Fuck no, I didn't pray! And I never saw any lights. Or any tunnels. None of that shit. The world just shrank down to a pinhole. My last thought wasn't 'Jesus Save Me!' Right before I blacked out, my last thought was: I'm gonna die *in the yard*? Seriously? After all the dumb shit that happens in the field!?"

Bobby Lee found Charley under the racks and used a gin truck to free him. "I owe my life to that man," Charley would say later with more than a hint of amused embarrassment. Once I'd catch him looking at Bobby with his forehead screwed up and one eyebrow raised as if to say, *Really?* At the time of the accident, Charley's heart did stop. Technically, he died. EMTs brought him back to life.

I BEGAN SWAMPING for Charley in December, when Smash got a job at a local machine shop. Our first rig move was on an unfinished location where scoria had been dumped across half the site.

Where the scoria ended there was a steep three-foot drop to hard prairie clay. Haul trucks barely had enough room to turn around. Traffic on and off location was a gnarled knot of cursed impatience. Making things worse, what scoria there was hadn't been packed down.

A bed truck, onto which was tied a 90,000-pound generator the size of a fire truck, was sinking into the rocks. Wheels spun for purchase as the truck's ass only slipped deeper. Tex sat behind the wheel with a blank expression on his face. Was it terror? Exhaust poured from the back of the truck as the wheels spun, rocks spitting in a dirty red arc from beneath the whirling tires, cables straining visibly under an immense amount of pressure. The generator, at any moment, could have slid off the bed and flipped the truck over.

Charley and I positioned the crane so he could line up his pick, but the outriggers were sinking into the scoria, too. We had to reset four times, and my arms were jelly from throwing the heavy wooden pads beneath the outriggers, then dragging them out and tossing them back onto the body of the crane. We got set for the fifth attempt, and I watched as Charley extended the boom. The crane's cab tottered forward, and my heart leaped into my throat. "Fuck this," I heard a hand say behind me. A group of us backed away.

"This is the dumbest fucking thing I've ever seen in my whole fucking life on this godforsaken fucking stretch of earth," Charley howled into the company radio with gleeful rage. "If I eat shit and die here, I'm going to haunt all of you stupid motherfuckers until every last well in the whole of this entire fucking Bakken runs dry. Then I'll fuck all of your daughters. As a ghost. I'll ghost fuck your daughters to death. And then your granddaughters. I don't give a fuuuuuuck!"

Charley pulled the boom up, and we leveled out the crane a final time. When he lowered the line again, I inched carefully toward the bed truck, wrapped slings around the generator's feet, then backed quickly away. Charley pulled up on the line and Tex gunned it, his jaw set, his eyes straight ahead. The generator slid off the bed just as the truck dragged itself out of the stones.

"I guess I won't have to do that now," Charley informed the crew over the radio, "and by 'that' I mean ghost fuck your whore daughters when I'm dead."

By mid-January, the three of us—Charley, Huck, and I—have become something of a crew. While Charley doesn't hang out with us much outside of work—he's got a wife and kids—we drive to and from location together, we crack each other up, we talk about things that are important to us, and we help each other out. Huck is often late, and in an act of generosity that never fails to surprise me, Charley directs me to pick him up when he oversleeps.

I would witness Charley obliterate field hands, decimate fools, and utterly destroy slow-witted truck drivers with a sharp tongue as unrelenting as a machine gun, but he never gave Huck too much of a hard time. He was a famous work gossip as well, but he never ratted Huck out to dispatch, either. Huck was always in danger of getting fired, and with his impending court dates, it was imperative that he hold on to his job. So, while Huck helped me learn the crane rigging, Charley covered for him.

"MY WIFE WAS trying to tell me we were 'meant to be together.' I said 'No, we fucking aren't,' and she got all offended." Charley sits jawing from the passenger seat as I drive us home after a day's work. "She's been talking to her friends who have all these sayings all over their house. It makes me want to vomit when we go there. 'Bless this House' and all that shit. I said, 'We aren't meant to be together. We fucking choose to be together. It's actually a lot fucking sweeter that way. If we were meant to be together, then we're basically in jail. And I've been in jail. I know all about jail.' Just like Huck is going to know about jail real soon, too."

Huck pipes up. "I just received a plea deal in the mail. They're offering me a one hundred and twenty day stint for spitting on the police officer and four thousand dollars in fines for damaging a police vehicle."

"Well," Charley begins, "that's a real fucking shit sandwich, but don't be so glum. You're probably lucky it isn't worse. Cops tend to dislike being spit on by dumb, drunken fucks. Even I never did anything that stupid. I did go to jail, oh, I don't know . . . six times, I think? But never for more than thirty days at a stretch. You should see if you can start serving it right away so you can dodge this winter. You don't want to get locked up as soon as it gets nice outside."

"Four months," says Huck. "That'll change a person."

We stop at a convenience store to pick up cans of Four Loko, a weird alcoholic energy drink, and the three of us sip them as we continue our trip back to the yard. They taste like Smarties.

"The Wildebeest says Charley is the smartest motherfucker in Williston," Huck tells us.

"Charley *is* the smartest motherfucker in Williston," I second, and Charley agrees. Later, however, when I think back to what actually made Charley seem so bright, I'll have trouble coming up with a good example. The primary outlet for his intellect was hurling insults. Woe to the man on the wrong end of his tongue-lashing! I saw Charley cut roughnecks to ribbons, and I was subjected myself, many times, to his obliterating onslaught of derogatory insults. There were times I fucking hated him, frankly, but I always had to hand it to him: Charley could talk shit like Picasso painted. Trying to explain that is like sketching *Guernica* on a napkin.

He'd been born in a pre-boom rural Williston without any kind of guidance or hope of getting a career he could wrap his brain around. In western North Dakota in the early 2000s, guys like Charley were looking through the obituaries to find jobs. "That guy died? Where did he work? I'll call them up!" If he had grown up somewhere else, it is easy for me to imagine Charley becoming a media personality. He listened constantly to talk radio. "If I did that, I could talk like a twelve-year-old for the rest of my life and get paid for it," he told me.

Charley's girlfriend got pregnant when he was fresh out of high school, and he took a job as a cook and a second job as a dishwasher. Rent was still too high for his wages, so he shoveled the walks for his landlord during the winter. "I was living the struggle," he said. "When I found out my daughter was coming, this, you know, sweet little bundle of joy, I was like, 'Awesome! I can't even take care of myself!'

"After she was born, I used her diaper bag on several occasions as a vehicle for stolen property. We couldn't afford formula and diapers and shit, so I'd take the diaper bag into the store and just stick like two cans of formula and a pack of diapers in there. What are they going to say? You're stealing that!? No, I'm not. I fucking carried this in here."

"What did you get locked up for?" I ask.

"Dumb shit, mostly. The first time I was so blacked out I didn't know what happened until I heard the judge read the report at the hearing. Apparently, I'd been out having drinks with friends. We ended up outside somebody's house in this, like a cul-de-sac. I walked into the neighbor's house. The door was unlocked, and I went upstairs into the bedroom and there was a young couple in bed. They're waking up like 'What's going on?' I said to the guy, 'Get the fuck out of here. I'm going to fuck your wife.' And the guy got up and left. He actually left me alone in the room with his wife! What a fucking pussy. I guess he went downstairs and called the cops. I didn't fuck his wife, though. I just wandered back out to the party. The cops showed up and tackled me from behind. I resisted so they hog-tied me and took me to lockup. I learned all this the next morning. I was still drunk. They read the charges and I was like, 'What the Everlasting Fuck?' It was like hearing about some dumb shit some other dumbass did, but it was me! By the time the judge got to the end of the charges, I was laughing my ass off. I would have gotten out on my own recognizance that day, if I could have kept my fucking mouth shut, but the judge told me I was making a mockery of his courtroom. They threw me back in my cell. I was an idiot. I was an angry young man."

He met the woman he would marry after he found out his girlfriend was pregnant and as he was headed toward a stint in jail. His future wife's parents hated him, of course. How could they not? But the two fell hard for each other, and she corresponded with him while he was locked up. Charley's wife would later tell me that her high school boyfriend was a prison guard at the jail where Charley was doing his time. "That was pretty embarrassing," she said, and laughed. Of Charley, she continued, "He got out of jail and decided to kind of straighten out his life. That's when he started dabbling in the oil field, and he went from being a troubled young man to finally having a job. He went from riding a bike to finally owning a car. Since then he's just done awesome building our life together and our family. When I think back to when I first met him, did I ever think that we'd be here? You know, four kids, a house, married? No. But I'm glad that he straightened up. Yeah."

SUN DOGS

The horizon is all shades of white. Snow falls on a windy slant and holds tight to the prairie's dirt crust, sparkling and reflecting. Ice grapples a million glittering stalks of grass rigidly waving and clinking against each other. The whole sky is a single cloud stretched thin.

Huck and I leave the crane pickup and walk into the insanity of cold. The very air sparkles as the moisture in the atmosphere freezes before our eyes in undulating shimmering waves. Huck dances forward and points toward the sun, raising its cold white head in the east. The sun looks tall and pinched, not like an orb but in the devilish shape of a goat's pupil. Circling the sun is a purely symmetrical crown of light.

"There they are," Huck says, "the sun dogs."

In the ring of the halo, on the same eye level as the sun, an arm's length to the left and right of the goat's pupil, burn two mock suns. The fiery outstretched hands of an ancient winter god inviting praise.

"Pretty cool, huh?" says Huck.

"It is pretty cool," I say.

It is negative 38 degrees in the Bakken. We get to work.

The day smears past. The soft sound of snow rustling on snow. The tinkle of frozen stalks of prairie grass. Generators and idling trucks. The rip of a wild wind through location peels the snow from the earth and sends it spiraling sideways through the machines, through the zippers of our work clothes. A flake of snow licking the neck feels like a cold fist. Chains rattle and gears shift. The sun dogs move higher in the white, stretching into day. Huck appears, big and commanding. His normal awkwardness gone, his body poised and moving. Legs, arms, and head in perfect tandem with the crane. He's throwing hooks, shortening chains, and racing after trucks.

No words are spoken in the silent din. I watch him as he kneels next to me. His mask covering his face except for the pink around the eyes. Icicles extend his eyelashes—delicate, otherworldly, feminine, as his eyebrows fill with ever-growing glacial mountains of ice. A flea's Antarctica. Huck rigs a piece of metal to the gin truck and disappears after it.

I CLIMB INTO the pickup to warm up. Pulling off my hood and mask, I feel the ice on my skin melting. Tiny rivulets of water run down my coat. Wet face, wet jacket. Blowing into hands, rubbing eyes. The company radio buzzes with trash talk. Rubbing hands together. Stomping feet. Closing eyes and breathing deep. Don't pull off your boots. And don't stay too long. Don't let the feet warm up. Get too warm and it only gets worse. If you let your feet sweat, the moisture freezes as soon as you're outside. Ice in your boot ain't just cold, it's mean. Unrelenting. Unforgiving. Unshakable. Cold as a burn. Like an impact driver to the foot. So, get out. But it's so warm in the truck. But get out. Just a moment longer. No, get out.

Back into the cold and moving into the white. Part of the machine. Part of the snow. A son of the sun dogs. Swinging chains and rigging hooks. Keeping legs moving. Keeping arms moving. Moving body across the hard earth. Crunch of boots. Crunch of metal where the black gold flows.

I stop for a moment. Just a moment. I close my eyes. It is a simple moment of rest as the crane idles by the substructure of the rig. I exhale through my mask, but when I try to reopen my eyes, I can't. My eyelashes are frozen shut, fused together by ice. My gloves are covered in invert and diesel. I can't touch them to my face. I pull off my right-hand glove with my left hand. Immediately, all feeling leaves the exposed hand. I can't open or close it. It sits at the end of my arm, a useless piece of meat. I paw at my eyes to brush the ice off my eyelashes. My hand as awkward a tool as a man could wield. But it works, I can see again. I blink my eyes free. I look at my hand. A meaty red paw. Hurriedly, I attempt to put my glove back on, but it won't fit. I can't will my fingers to conform to the fingers of the glove. I try again, but I'm trying too fast. My fingers won't move, they won't slide into the glove. I slow my breath, I fight my panic, and I move deliberately, slowly fitting the glove around the foreign hand. I get it on and then bury my hand under my armpit to warm it. I hear a horn. The crane swings toward the north end of location. I move after it.

AT THE END of the day, in the final bitter hour of strangled light, Huck and I rig down the crane. We'd passed each other all day long, jumping onto each other's rigging when we could. Throwing hooks and chains on each other's lifts. But never talking. Just moving. On from rigging and back into the white and then on to rigging in a silent dance of brothers.

"Well, you can go home tonight and know you are one of the toughest motherfuckers on the planet," Huck finally says to me. "Ain't just anybody, barely anybody on earth has worked a day like we just did. That's something to be proud of."

We pile into the truck to drive home. Charley is already running his mouth, fiddling with the satellite radio and popping open a Four Loko. Me in the driver's seat backing up the truck, peeling out onto the road, preparing to be made fun of for my driving. Huck spreading out in the back seat slipping a dollop of dip behind his lip.

I knew then that it had been a special day. But it wasn't until several years later, after Huck died, that I would remember it, somehow, as one of the best days of my life.

SUPPER

January continues to crawl past at a truly glacial pace. It is another brain freeze of a morning when the Gruff Crane Operator picks me up for work. We grunt in greeting and I hop into the cab. He pulls the truck onto a main road, and we immediately hit a patch of ice. The rear wheels swing wildly off to the right and then left, fishtailing for purchase. The crane operator aims the truck in the direction we are already barreling, so the wheels can catch some gravel. The instant they do, he slams the wheel back, centering us on the road a split second before we would have slammed into a ditch.

"Lil slick out," he mumbles.

On the hour-long drive to location, we pass a jackknifed tractor trailer twisted up on the shoulder, a Chevy flipped upside down in the middle of a median, and a Ford smashed in a ditch. Another winter day, another drive in North Dakota.

Approaching the job site, we see a row of pickup trucks idling on the side of the road outside the lease. Something is wrong.

"Tool pusher's shack burned down last night," the lead driller tells us. "There was a natural gas leak by the generator. It caught late at night, and we all got up and put the fire out. Well, we thought it was out anyway. It caught fire again this morning. We tried to put it out, but gave up and let it burn. Pusher got out all right, though. Nobody got hurt."

The gas leak occurred right by location's designated smoking area. It was a stroke of luck that nobody flicked a butt and blew themselves up the day before. The shack is a burned-out shell of twisted, blackened metal. It's been dragged away from the other trailers and sits forlornly off by itself.

A small construction crew works near the source of the leak, using Bobcats to excavate the area, locate the break, and hopefully patch it. Until it is fixed, we aren't allowed on-site. We aren't allowed to go home just yet, either.

The Gruff Crane Operator pulls the truck to the side of the road, and we sit there, just off location. It is warm and cozy in the cab. I sip a watery coffee and snuggle down into my seat, pulling my jacket up around me and falling in and out of sleep. A glorious North Dakota sunrise fills the window to my right. If I'd grown up here it would be possible to take it for granted, I think, but I wouldn't know how. Fluffy peach-colored clouds float across a radiant turquoise canvas. It looks like a doctored photo or an image from a sci-fi movie. But so very real. The pastels shift before my eyes.

After an hour or two, the Gruff Crane Operator digs up a portable DVD player and we watch *Live Free or Die Hard* in the truck. We while away the whole day there, watching movies, talking a little bit, occasionally stepping out into the cold to piss or gossip with the other hands. It takes all day for the construction crew to locate the leak, and at that point we drive home. All in all, it is as good a day as the oil field ever begat.

As we head back toward the shop, the Gruff Crane Operator is talking on the phone with his wife. He turns to me. "You like cube steak and hash browns?"

"Yeah sure," I say.

"Well," he returns to the phone, "I guess Magic Mike just invited himself over for supper."

The Gruff Crane Operator's name is Jonathon. His wife is Lisa. Their apartment is small and cluttered. I walk in and toss my jacket on a chair. Jonathon pours me a vodka and Sprite, and we join a small gathering in the living room. Lisa teaches at the local high school. The school's principal and another teacher sip drinks on a puffy couch in the center of the room. Jonathon takes a seat in a big oversized recliner, and I settle into a chair. Jonathon and Lisa's adopted son, a nine-year-old black boy, runs around the room playing with a fake bow and arrow.

"What you got there? Picking off a doe?" Jonathon asks his son.

"A buck," the boy tells him.

Jonathon lets out a hearty laugh. "He's gonna be the biggest ole redneck you ever saw." Pride swells his eyes.

It's a little awkward to be among this tight-knit group of locals. I find myself tongue-tied. They talk about the challenges of teaching during the boom. The school system is flooded with kids from all over the country, all over the world. Some of these new students don't speak English, and there are very few teachers who can accommodate them. And the kids come and go, their parents chasing jobs, as they chase schools. Some classes start with fifty kids and end with seventeen. The teachers do what they can.

They talk a bit about the upcoming play, and I tell them how much theater helped me get through school. We put on some of the strangest theater pieces a high school has ever seen, including an African version of the Oedipus myth called *The Gods Are Not to Blame* and a play about the Holocaust called *Roses*. For the stage set on *Roses*, my teacher Carl, who'd written it, strung barbed wire between the audience and the actors, dumped hundreds of pounds of dirt on the stage, and created costumes made from burlap sacks. It was an incredible experience. Photos from that play are shocking.

The high school in Williston is mounting a production of *The Music Man*. Lisa says I should volunteer. "We could use the help," she tells me, and the thought fills me with a wonderful, inspired feeling. I could see myself doing something like that, helping the kids, giving back something like my high school teacher had given me. The conversation moves on, but I keep turning the idea over in my head. Of course, I couldn't volunteer at the school. I don't have time like that. It would get in the way of work. It makes me wonder what any of it is all about.

THE NEXT MORNING, the job site is crawling with safety men. Clean hard-hatted Texans parade proudly around the perimeter sipping coffee and pretending they know what they are doing. Before work starts, we crowd onto the drilling floor for a meeting. Company higher-ups have driven in from Minot and Dickinson, others have flown in from Texas. The floor is packed tight. There is an air of excitement.

Once he gets our attention, the tool pusher discusses the gas leak. Somebody makes a crack about how good he looked using a fire extinguisher in his underwear. He grins crooked and comments that he knew something was wrong when he realized the temperature inside his shack was actually comfortable. Then he drops his smile and talks about the safety issues on the rig. "I don't want anyone to get hurt," he tells us. It is one of the only times that I've gotten a sense of sincerity from a boss about safety since I started. The tool pusher lets us know that men from the company have arrived to keep us informed on the status of the leak. He introduces one of his bosses.

His boss has the look of a teenager on a field trip, his face an expression of perpetual amusement. He's middle-aged, plump, and unworried. Sunlight hits his hard hat, revealing nary nick nor scratch. He clears his throat. We watch him, and even among the hardened men, a silence falls. I get a sense in that moment, that aside from the tool pusher's wisecracks—it was his accident, his house that burned down, he who could have been caught in that fire and killed; he's earned the right to be casual about it, to brush it off, to joke—that the gathered men are serious. We want reassurance. We put our lives on the line every day. We do it for money. For families. For dreams. To pay off trucks. To buy houses. To put kids in schools. And we get paid well, but really not that well. And if we stop, the world stops: no one flies, no one drives, no one drinks from a plastic bottle, eats off a finished tabletop, or swallows a gelatin pill. We do that. And at what cost? We wreck our bodies and risk our lives, then we laugh off the danger because we need to. We need to be *bad motherfuckers* because the alternative is fear, and fear fucks up. It makes mistakes, and mistakes get people hurt and killed. There was a gas leak on location *in the designated smoking area*. That scares the hell out of me. I want to know I will be taken care of. We look at the man, but the man doesn't look at us. "If I were you," he says, "I'd be nervous."

OLHOUSER'S FAREWELL

Huck appears in the inky black syrup of morning, standing in the yard with a cigarette in his mouth. The previous day, during a move, a haul truck driver slid off the icy road, landing his rig in a ditch. As they were trying to unfuck the situation, Olhouser took a spill.

"I thought he retired?" I say.

The Viking had told us a few weeks earlier. It was after a long workday, and a local farm boy swamper offered Huck and me a couple beers out of the back of his truck. Snow sat on the ground in the yard, and while we were talking, the farm boy scooped some up into his hands. He tossed a pile into the air, and let it fall onto his upturned face.

"You definitely are a local," Huck said.

That's when the Viking pulled his truck around. He wasn't wearing a hard hat or a ball cap, and his blond hair cascaded over his broad shoulders. "What's up, you Thor-looking motherfucker?" Huck asked.

The Viking rolled down his window and gave us the news. It was like hearing Davy Crockett had moved to China or Superman was returning to Krypton. Olhouser was retiring, a minor oil field deity stepping down from a grease-soaked Olympus.

We talked it over for a while, each of us letting this new reality sink in. The Viking did his Olhouser impression. "Oh, you guys are fucking me!" he said in a high-pitched squeal. We shared stories of Olhouser kicking his hard hat off the top of a rig—it flew straight off into the horizon, chasing the flat line of prairie all the way into Saskatchewan, where some say it impregnated a farmer's daughter. We talked about Olhouser's injuries. He'd been slammed in the head by a shaker tank, and he broke both legs when he fell off a drilling floor one time. I did my impression of Olhouser, imagining him after the accident, pulling himself around on his hands and elbows, furiously trying to get everybody back to work. "Get the fuck back over here and make a fucking hand!" I hollered.

"He hates to see the trucks idle," the Viking said, nodding cheerfully. "He just can't fucking stand to watch them sit still. Drives him nuts! I'll miss him when he retires, though. I will."

"I've only been in the field for a few months, but I feel like I'll miss him too," said the farm boy.

Huck and I grunted our approval. "Oh yeaaaaaah." It was strange to think Olhouser wouldn't be around.

"THAT CRAZY OLD MAN will never retire," Charley says, crunching across the gravel toward Huck and me.

Olhouser had banged his head when he fell down, and he was knocked out cold. "When he got up he didn't know where he was," Huck tells us, visibly shaken by what he'd witnessed.

"Did he go sit in the van or something?" I ask.

"No. He went right back to work." Huck looks at Charley and me, and we share a laugh.

It wasn't the first time I'd seen the old truck pusher experience a close call. A month or so previously, I was standing with Olhouser and a tool pusher on the driver's side of a haul truck, unloading a mud pump. It was twilight and becoming difficult to see. The pump wasn't centered correctly. It kept getting hung up on the steel pins at the end of the trailer, giving off a terrible screeching sound. We had to keep moving the pins around to guide it.

At Olhouser's command, the driver loosened the bridle, and the mud pump lurched back. There was a loud *crack!* like a small-caliber rifle popping off, and Olhouser looked at me, eyes wide, mouth agape. He leaped straight up in the air and started hooting, "Whoo-whoo-whooo!" It wasn't that unusual—the old man could get excited during rig ups, so I just smiled at him. Then I saw a big chunk of metal drop onto the mat at his feet. One of the pins from the trailer—steel eight inches long and two inches wide—had snapped in half and flown directly at Olhouser, landing in the crook of his elbow.

"Jesus Christ!" hollered the truck pusher. "Are you okay?"

Olhouser stopped hopping and hooting and peeled off one of his trademark high-pitched cackles. "Oh, I'm fine," he said, "I'm fine."

"Are you sure?" Another hand joined our group, and we huddled around the old man.

"Oh yeah, I'm fine. I'm fine." Olhouser started swatting us away.

"Do you want to take a break?"

"No, no, no, no," said Olhouser. We were ruining his fun with concern, and he couldn't stand the attention. The tool pusher and the other hand stepped back and Olhouser looked at me, his eyebrows raised. "Right in the crook of my arm!" he shouted. In a flash, he was trudging off to the next task.

I exchanged a look with the tool pusher. "I think I'd want to sit down for a few minutes and lower my heart rate if that happened to me," I said.

"I think I'd go sit in the trailer and cry," the tool pusher responded.

I'LL SEE OLHOUSER one more time before leaving North Dakota. He appears in front of me shouting about blizzards hammering the East Coast. He seems disappointed we're not there getting clobbered by snow. I ask him if he's seen a doctor since he got knocked out. "Oh no, no, no," he says. "I just need to get to the chiropractor."

"It sounds like you got a pretty serious concussion," I say. "You probably want to go to a medical doctor, right?"

"Oh no, no," he says, warding me off like a nagging wife. "I've already made an appointment!"

He leaps into his truck and it bounces off. "I love that lunatic," I say out loud to no one in particular. After thirty-eight years on the job, a man who worked with the joyful clamor of a thunderstorm departed as quiet as a mouse rustling through prairie grass.

A SERIES OF CLOSE CALLS

Y ou hear about that roughneck out on Ensign 166?" Charley
asks me.

"I don't think so."

"Down near Watford? They were drilling west out toward the
badlands down there."

"Uh-uh."

As a rule, Charley doesn't leave much unsaid. But he pauses
here. "It's weird when it happens to a guy you know," he tells me.
I'm driving. I glance over at him sitting swamp side. His expression
buried beneath his shades and beard, there's a trace of searching in
his voice. "The guy working with him was standing below him on
the floor of the rig. When it happened, he said there was just a wall
of blood that poured down in front of his eyes."

"Jesus Christ."

"He knew right away that they'd lost him."

"Damn," I say.

"Yeah," Charley responds. "It's just kinda weird when it's a guy
you know."

A FEW DAYS LATER, I'm working with an old-timer gin truck driver with a bushy mustache and hair white as copy paper. I've worked with him plenty of times now. He is ornery, eccentric, and careless. "Fecal matter!" he shouts when he makes a mistake. The last time I swamped for him, he broke a pipe fitting off a tank in the rain and flooded the tank with antifreeze. The first time I'd worked with him, he smashed into the Company Man's pickup truck. On this move, he'll nearly kill me.

The gin truck lifts are still in my bones. When I hop out the door onto the dirt, I walk behind the truck and throw the chains without thinking. Then I watch the picks raise level. One after another after another. Easy as pie. When all is said and done, let it be known: the Wildebeest taught me well.

The pump is sitting on the back of the haul truck as part of a shit load. The pick is small in oil field terms; it is about the size of a lazy chair and weighs probably half a ton. Ole Bushy, operating the gin truck, backs up to it, and I climb onto the trailer. He lowers the chains. I grab the hooks and link them to the pump quickly, but just as I do, I see that a pipe fitting on the pick is entangled in a second pump—about the same size as the first—sitting next to it on the trailer. I need to untangle it so I raise my fist to signal the old-timer to hold position, but he doesn't. The poles on the truck begin to raise, picking up the slack of the chains. My back is to him. I signal again, silently, quickly with a closed fist, but he keeps pulling up. I'm standing between the two pumps and the edge of the trailer. I swing my head around to see that Ole Bushy isn't looking at me at all. Time slows down.

"Hold the fuck on!" I yell, but he doesn't hear me. "Boom down!"

I holler again, louder. The chains grow taut and the pumps shift at my feet, dragging toward the trailer's edge. One catches my boot, and I nearly lose my balance. I start sliding toward the edge. It's not a long drop to the scoria but the ground is frozen solid and I'm in an awkward position. If I fall, I'll land on my back and the pump will drop directly on top of me, all 1,000 pounds. I yell again, this time as loud as I can. "Goddammit, hold on!" I've got one hand on the chains, my other hand gesturing wildly for him to stop. The old man is oblivious.

I'm inches from the trailer's edge when I pull my boot free and push myself awkwardly away from the pick just a moment before I'm knocked off balance. I barely manage to stay on the trailer, slipping and landing hard on my knees. Wincing, I turn to watch the pumps, a tangled mess of metal, hoisted up into the air. They dangle there for but a moment, silently rotating in the negative space between the truck and the trailer. Then the second pump slips free. A thousand pounds slam into the frozen earth with a dull thud. If I had moved a moment later, I would have been under it.

LATER THAT DAY we have to pull a generator off a haul truck. The generator weighs about the same as a small house. It requires an assisted lift. When I climb onto the trailer, I see the haul truck's bridle is dogged out, or slack. I connect the gin truck's slings to the generator but signal Ole Bushy to wait until the haul truck driver puts some tension on his line. I then signal him to drop his number two, lower the poles so his line will be centered directly over the pick. Otherwise, the whole thing can swing off and flip the trailer. Like the Wildebeest had told me one of the first times we did this move, "You get sick of carrying dead bodies off this motherfucker."

Again, Ole Bushy doesn't heed my signals. I'm straddling the haul truck's bridle line when he pulls up on his winch. The generator is yanked a few feet into the air, and it slips a good foot and a half toward the swamp side of the trailer. I feel the whole world shift under me. I trip over the bridle, stumble across the sliding trailer, and drop down into the scoria on my hands and knees. The generator swings above my head unsteadily, straining the line. I scurry quickly out of the way. I'm lucky. If the pick had been off by another twelve inches, it would have sent me flying, and if the cable had snapped, it would have cut my legs off.

I could scream at the driver. I could tell him he's a dumb fuck and he better fuckin' watch it or I'll bang his fuckin' head in. I could stomp over to Bobby Lee and tell him I ain't working with the old coot anymore, so he can put me with somebody else or I'll go sleep in the fuckin' van. I've earned my right to make those kinds of demands. Nobody would blanch. It would just give everybody something to talk about. "Ole Magic got fuckin' pissed, bro!" But I don't do any of those things. I don't feel any anger at all. I feel acceptance.

LINKS IN THE CHAIN

One morning, on location, I receive a message from a friend telling me that Pete Seeger has passed away. I corresponded with Pete in the early 2000s. He was a friend of Woody Guthrie's, but way more square. As he told me, "I don't drink. I don't smoke. And I don't like singing in nightclubs!" We traded letters and talked on the phone a few times. Pete was generous with me and a lot of other people, too.

He was an oak tree of a man, and he had lived many lives. Pete assisted Alan Lomax in putting some of the very first recordings of American folk music on acetate disks. He learned to play the banjo along the way, and he used music not just to tell stories but also to organize people, to protest, to instruct, to radicalize, and empower. It is one of the reasons I've been drawn to folk music myself. This idea that it does something. It has a use outside a concert venue. It can take you places.

Pete held the picket line with striking workers in the 1940s. He served honorably during WWII, entertaining the troops with his banjo. He came home and built his house in the Hudson Valley with his own hands. He had a hit record out when he was forced before Congress's House of Un-American Activities Committee.

"I have sung for Americans of every political persuasion," Pete told the committee, "and I am proud that I never refuse to sing to an audience, no matter what religion or color of their skin, or situation in life. I have sung in hobo jungles, and I have sung for the Rockefellers, and I am proud that I have never refused to sing for anybody. That is the only answer I can give . . ."

His passport was confiscated. He faced hard jail time. Blacklisted from TV or club work, Pete took his music to college students and children. He later marched with Dr. Martin Luther King. When I was researching Woody's life, at his official archives, then in New York City, I saw the telegram King had sent Pete inviting him to the March on Montgomery. If objects can hold magic, that one does. The old piece of paper took my breath away.

Pete never stopped. He built a sailboat called *Clearwater* and championed efforts to clean up the Hudson River. In 2008, he played "This Land Is Your Land" with Bruce Springsteen on the Capitol steps at the inauguration of America's first black president. He was a giant who strode across the American Century, a good-hearted, banjo-playing colossus.

He also spoke patiently over the phone with me, singing songs into the receiver as they popped into his head, commenting on the hummingbirds outside his window, and expressing concern that the world wouldn't be around much longer. He assured me that if it was, people would still "be dusting off Woody's children's songs and singing them for the kids," he said. "So simple."

On the frigid plains of North Dakota, stomping across the frozen earth and swinging chains around steel, I think about the conversations I had with Pete, about the letters we exchanged. Our correspondence is a small fire in the narrative of my life, a memory I return to for warmth.

Throughout my life theater had connected me to my mortality. I thought about this when I heard of Pete's passing. In theater, you spend three weeks in an empty room digging deep inside yourself, connecting with other actors and connecting to your own soul through the words and actions of a character written by a playwright. Then for a week, walls are raised around you. The empty space is filled by a stage set, a drawing room maybe, a country house, or something mechanical and abstract. You are dressed in a costume, lights dance in your eyes, and music plays. When the audience arrives, you feel the butterflies of birth. They observe as you enact a ritual of words and movement every evening at 8:00 p.m. for a weekend, or four weeks, or many months. And when it is over, the set is torn down, the lights are packed up, the costume is returned to the rack. The words and movements that connected you to the other actors and to your own soul may linger in your mind for years, but they are never to be performed again, not in the exact same way. Eventually, they are forgotten. You have a wrap party with the cast, a group that can feel as close as a family, and then, as often as not, you never see them again. You grieve. It is an experience as joyous as life and as sudden as death. In a strange way, it is a lot like working in the oil field.

Playing folk songs, like Pete did and like I do, ties you to eternity. Studying and singing a song such as "Saint James Infirmary" is to feel a living thing like an ancient river move through you. The words and the melody to a song like that have a lifespan longer than a redwood: before they were sung by Pete Seeger, before they were sung by hillbillies and slaves, these songs were sung by Celtic farmers, by Vikings and African griots in forms and with words we might still recognize today. When these songs are sung well, I believe you can hear those voices and that eternity joining in and singing along.

True actors and musicians live for the transcendent moments where everything disappears but the action. The audience, the microphones, the stage set, the lighting; a person's own vanity evaporates, time recedes, and only the saying of lines, fretting of chords, or singing of melody remains. As weird as it sounds, when working in the oil field, I experienced this sensation with more frequency and at greater duration than at any other time in my life. I got lost in the work, and it was prayerful. I'd found meaning here, grace.

I tie two hooks around a greasy piece of heavy iron, linking them back into themselves and then shortening the chains carefully. I signal the crane and the metal starts to raise but I don't like the look of it. I signal the operator to stop and then come down. He eases the load back onto the ground. Carefully, I count out nine links of chain, unhook the second hook, and place it where I counted. I signal the operator again, and the crane raises the metal. This time: level as a level. I wave and the operator flies it out.

Folk musicians can be referred to as links in a chain. The chain is the music, and the singers are the links that allow the tradition to continue. Simply put, the songs are passed from one person to another. More deeply considered, this philosophy is the opposite of rock-and-roll or pop music. In pop music, a song is used by a singer to express themselves. In folk music, the singer is used to express the song. To be a folk singer is to make yourself a tool of tradition, to become a physical piece of it, a link in a chain. To sing folk music is to serve, and the life you build by doing—your very flesh and blood and what you do with it in the time you've got—is as important as the music itself.

I admire these guys—Woody Guthrie, Pete Seeger, Ramblin' Jack Elliott—but not always because I love the sound of them. Recordings of Woody can be scratchy, hard to listen to, and Pete sometimes comes off like a kindergarten teacher. Jack is one of my favorite vocalists of all time, but he'll go off-key and stubbornly stay there for entire concerts. I admire these men because of the way they each live or lived truly singular lives, wrenching life to suit them, molding it to the mud of their own passions and preoccupations, giving voice to something bigger than themselves, telling stories of the people we don't hear much about, forgotten people, dispossessed people, homeless people, immigrants, migrants, refugees, beat-down farmers and wounded soldiers, busted cattle rustlers, people segregated by the color of their skin and the weight of their wallet, the castoffs, the no-good nobodies, the losers. Because, with these songs these people are raised up, ennobled. The songs raise all of us up.

"Yo, motherfucker!" The hollered words pull me out of my reverie. I look up to see a swamper struggling on the ice with a mat. "Gimme a hand with this thing," he says. I rush to him, stomping my feet hard into the ground to break up the ice and give myself a little more traction. I get my hands on the mat. We right it and quickly land it on an idling haul truck.

The swamper rubs his gloves on his knees, then turns to me. We're both covered head to toe in winter gear, masks cover our mouths. All I can see of him is the ice in his eyebrows, the lines around his eyes, and his eyes themselves, vividly bright in the frigid cold. He looks me square in the face.

"Thanks, Magic," he says.

JANUARY'S FINAL COLD, windy, icy day. Rushing across location with the sun glinting off the snow broth. Moving my legs and arms and moving the rigs. Wind cutting through the zippers of our jackets. Stained and streaked in grease and invert. Someone needs to climb the hopper tank.

The last time I did it without a harness. This time, no thanks. I find an old harness buried in a compartment of the crane, and I put it on. Too big for me, of course, but better than nothing. Walking up to the ladder, I pause to look up toward the top of the hopper tank. The crane is tied onto it, some 70 feet above my head. I click the carabiner around the ladder and slowly, big arctic boot step after big arctic boot step, begin my ascent. Below me, Bobby Lee keys the radio and offers calm and cool instruction to the operator. Every few steps up, I disconnect my carabiner and reconnect it higher, looping one arm completely around a rung. If I slip, I might break my arm but I won't fall.

As I ascend, I don't look down or contemplate the ground. I keep my eyes forward, focused on the ladder and the carabiner that tethers me to it. When I do think of the ground, I feel my stomach hollow out, and a slight squirt of adrenaline spills into my blood. The feeling is good.

At the top of the hopper tank, high above the plateau that surrounds the job site, I loop my elbows around each rung and crawl toward the shackles in the D-rings. Once there, I pause and look around me. From this vantage point, the land in every direction stretches on forever. It does not end, it simply fades into the sky, wrapping around me like a great big, colorless bubble. Beneath me, the men and their machines. Above me, nothing at all. I'm struck with the sudden strange feeling that if I slipped, I would go upward, into the sky, falling forever.

My gloved hands unscrew the shackles. My breath pushes through the mask on my face and shimmers in the air in front of me. A new feeling sparks somewhere deep inside me, somewhere murky and old, a hollow place a tender thread runs through. I feel the extent of the bubble around me and the depth of the hollow inside me. I am outward and inward at the same time, in the same breath. Like the links in the chain of a folk song, my breath comes from and moves toward the eternal. I am aching hands and arms, made of meat and blood, and yet as immaterial as this whole, big globe. The line between life and death is crepe paper thin. It is time for me to leave North Dakota.

I WOULDN'T LEAVE simply for fear of dying. It wasn't just that the work became too dangerous. It had been dangerous the whole time. But it was true that I'd begun to see my death in North Dakota too clearly, and bleeding out on a frozen slab of land hundreds of miles from the closest hospital was not, by any means, the way I hoped to go.

I hadn't planned on staying so long to begin with. I thought I'd work for a summer and then head back to New York with my pockets stuffed full of cash. When I realized that wasn't going to happen, I was still a worm. Stubbornness maybe more than anything kept me in town. Then I truly fell in love with the work. I fell in love with my new friends. And I fell in love with the life itself. After that, for a period of time, I could envision myself becoming a gin truck driver, then maybe a crane operator, and then a truck pusher. I could see myself at the center of the wheel, conducting the oil song. It would take time, I knew that, but over the years, I could work toward that goal. For some time, I became enamored with that vision.

ON MY LAST trip to New York, I sat in one of my favorite watering holes, a place called The Drink, at a bar top made from a thick slab of Brazilian teak that had once constituted the hull of a Maine sailboat. During Superstorm Sandy, The Drink was the main hub for the Rockaways relief work I participated in, and during that period, I had grown close with the bar's owner, a painter, named Adam, who for years had worked in the demolition business. We'd sat at that very bar top on many an early morning to organize construction crews, and inevitably we ended our evenings there, too, propped up on bar stools, swilling beers off the teak and telling stories. Adam had the grizzled countenance of a working man, the tortured sensitivity of an artist, and the hard-knock wisdom of a heavy drinker.

I'm in the middle of telling him a rig up hand story when he interrupts me. "Don't forget you're an artist, Mike," he says.

I'm surprised by this, but I brush him off, take a long swallow, and swing around in my seat to face the barroom. I'd slung guitar and stomped out rhythm on those floorboards dozens of times over the years, giving some of the best performances of my life to audiences of forty or fifty people in front of a small upright piano under a thrift-store-bought painting of a tall schooner cascading through a rowdy sea.

"*Artist?*" I snorted. "I've never made any money at that. I don't know what the fuck I am."

But now, after nine months in the oil field, I was much closer to myself—to the bone of who I really was. And while I knew I was an oil field hand, that I had maybe even become a good hand, at least on some days, I also found myself preoccupied by Adam's words, because I knew, from the sparkle in the eyes of people who couldn't be farther from the oil field than if they were on Mars, that the place in this world where I could most be of service was not in a lifelong career as an oil man.

Somebody has to tell the tale.

BOOMTOWN EARTH

Alone in the crane pickup with the driver's seat kicked back, staring through the snow-crusted windshield at the derrick slowly rising over the well. The wind across the prairie is strong and steady. As it raises, flags atop the rig's mast snap in the cold northern gale. Winter has ground into mid-February. I'm at the end of what will be one of my last rig moves.

It is warm in the truck. The heat cranked, my jacket and mask lie on the seat beside me. The ice in my hair melts and wets my head. I'll have to hop out soon enough and return to work, but the crane is currently idle. There is nothing for me to do at the moment but stay warm. I catch a look at myself in the truck's mirror—unshaven, sunken-eyed, red-faced, older and younger than I've ever been. I shrug. *I look good*, I think.

The derrick continues its arc upward. It is an image I have seen dozens of times since that day on Sidewinder Canebrake 103, over nine months ago when I started this work. Nine months—long enough to birth a baby. Maybe I should stick around. No, it is time to go.

I watch the men scurry around the rig in hard hats, jumpsuits, and safety glasses. Some of the nameless now have names, some of the anonymous are now storied. Some of these men are my friends. Even though I'm not a lifer, all of these men are my tribe.

I'm leaving regardless. I'm going off to someplace else. Someplace where I won't belong all over again. That seems to be my curse, a vagary of the blood I carry, a reckoning I'm bound to contend with over and over again. I thought for some time it was a search for a home. Then I believed it was a refusal of home, a tossing off of the very idea of home. Now what is it?

In nine months, I have become a decent hand. But I have learned what it means to be a good hand, and that is more important. A good hand shows up early. He is present. He listens. A good hand carries the heaviest load every time, takes on the dirtiest, most difficult task and doesn't complain. A good hand makes the hands around him better. No one is a good hand all the time. You have to make a hand every day. A hand knows his place in the system. He does his part to ensure that the whole machine runs smoothly. In a society organized around consumption, a good hand creates.

And what does that mean when I get out of here? How does this apply to the world outside of oil field locations? To a political and social arena that only grows more divisive and divided? To an economic landscape that grows less equal by the day? In a land where consumption by the individual is exalted while the health of the planet is essentially given only lip service? How does a person in this age of gross overindulgence become and stay a good hand? An ideal that I see requires being of the world but also fashioning that world into something better, bigger, greater than the self. In a society that worships leisure, how do you maintain the fortitude to get up every day and go to work?

Had Meriwether Lewis stayed on as Thomas Jefferson's secretary, he probably would have led an important life assisting the president in the new nation's capital. But he wouldn't have made a hand. Commanding the Corps of Discovery as it cut its way across this continent, through North Dakota to the Pacific and back again, that turned him and William Clark into good hands. Theodore Roosevelt could have lived the life of a gentleman businessman, and he would have been fine, like his father, a powerful man. Instead, after grieving and ranching in North Dakota, having slept in the saddle during roundup and brushed icicles out of his mustache while watching the cattle die, he returned to New York City and dove into the gritty politics of his age. Bare-knuckled with teeth snapping, Roosevelt entered the fabled arena. He made a hand.

Woody Guthrie could have weathered the dust storms, stuck it out in Okemah, Oklahoma, and lived the quiet life of an eccentric sign painter. Pete Seeger could have stepped before Congress, taken the Fifth Amendment, and returned to playing shows at the Rainbow Room in Manhattan. Joe Ehrmann, the former Baltimore Colt turned pastor who speaks so eloquently about the father wound, could have dedicated his life to coaching football games. He knows how to win them. I'd call my sister Kate a good hand. She's a nurse now. It's a high ideal but not an exclusive club. A good hand is a person who does honest work to the best of their ability every day and who offers that work to the world as a living prayer. The good hand is a servant. In the patch, the hand serves the petrol that greases the axis on which the planet revolves. But now? Outside of that? For me? With this hard knowledge I have gleaned through the arduous toil of golden and frozen days? With this coin I found? What will I do?

The sky behind the derrick is pale, exhibiting a kind of absence of color. An oil tanker cruises through the work site, circling the rig, the ground frozen so solid it doesn't kick up any dust. I open a protein bar, turn on the radio, and observe.

. . .

PITHOLE WASN'T CALLED "Pithole" because it was considered a dump by the people who founded it. The name wasn't a roughneck's joke. "Pit Hole" preceded the boom. Years before oil was struck at Drake's Well, years before Europeans settled western Pennsylvania, and centuries before the Seneca Indians made the region their home, somewhere between 200 BC and AD 800, an unknown group of nomadic people dug trenches, pits up to eight feet wide and twelve feet deep, into the soil by a small river in the Western Alleghenies. Cradled by timber, these pits were used to collect the grease that bubbled up freely to the surface of the earth, unaided by man-made derricks. Archaeologists can only speculate as to why they did it, what they used the grease for, but we know they used it for something. Perhaps magic.

OIL IS PROMETHEUS'S second gift to mankind, refined petroleum comparable only to the mastery of fire in terms of the impact it has had on the human race. And if the mastery of fire and the advent of language mark man's separation from the beasts of the field, then the mastery of oil is the demarcating line between modern man and all those who came before. Oil has, over the past 150 years, reshaped our lives in ways unimaginable to the 2.8 million years of human life that preceded us. It has reshaped our planet.

A few days earlier, Huck's brother-in-law said to me, after hearing an environmentalist on the radio, "Guys come out here and they sacrifice their bodies to make a good living. I don't see that that is so different from sacrificing some land so the whole world can have a good life."

In the years after I leave Williston, wildfires will devour millions of acres of forest not only across California, Australia, and Brazil but also in the arctic landscape of Siberia and Alaska. Droughts will consume large swaths of South Africa and Central America, leading to huge immigration crises. Hurricanes will hammer Puerto Rico, Houston, and the Bahamas, and floods will swamp the Midwest, destroying millions of acres of farmland.

In 2015, President Obama will lift the ban on US oil exports, making the United States the largest exporter of oil in the world. Meanwhile, an island of plastic garbage the size of Mexico floats off the coast of California, and the 2010s prove to be the hottest decade ever recorded.

The world that Lewis and Clark traveled as they followed the Little Muddy River through what is now modern-day Williston is gone. The teeming life of the plains they observed, "covered with buffaloe & buffaloe calfs, Elk deer &.c." is gone. The views of "large gangs on the opposite Shore I think we Saw at one view nearly one thousand animels" are gone. Buffalo skinners took some of it with single-shot rifles and railroad trains. Oil took the rest.

Every boom does bust, of course. It's a fact of life. The North Dakota boom isn't the one that we need to worry about. Boomtown Earth is busting.

IN THE CRANE PICKUP, in the frozen February afternoon, at the end of a long day that started the previous June, I chew on my protein bar and gaze out the window. I'm surprised when Waylon Jennings comes over the radio, a simple two-chord guitar riff over a four-on-the-floor bass drum stomp. It travels through my speakers via satellite, and when his voice enters the mix, thick as honey, smooth as dog hide and certain as science, I turn it up:

> Lord, it's the same old tune, fiddle and guitar
> Where do we take it from here?
> Rhinestone suits and new shiny cars
> It's been the same way for years
> We need a change

The rising derrick, like the ticking hand of a clock, moves irreversibly forward, from nine to ten to eleven o'clock. Time is funny here. In a place where the workday stretches fourteen hours. With a sun that in the summer just won't quit the sky and in the winter will barely make a day. Nine months is a lifetime, a thousand years a speck. The good hand of the clock turns toward completion, the good hand of the man turns up the volume:

> Somebody told me, when I came to Nashville
> "Son, you finally got it made"
> Old Hank made a hit, and we're all sure that you will
> But I don't think Hank done it this way,
> No, I don't think Hank done it this way

Winding its way easy as a tumbling stream over stones of melody, the electric guitar kicks in. On the oil lease, the derrick continues its arc upward, ticking ever closer to its destination. I settle into my seat, chew my protein bar, and take it all in. In moments it will be midnight.

SUNSETS

T hick, pink ribbons of cloud stretch the day's endless horizon above the final burning embers of a failing sun. I watch another spectacular winter sunset out the windows of the work van. Bobby Lee is driving, his ragged old cowboy hat pushed high up on his head. He's leaning back in his seat, sipping on a USA Gold cigarette and chatting so softly that I'm forced to lean in to hear him from the passenger seat.

"It's a devil's bargain," he says, speaking of the boom. "It's a good way to make a living but it'll ruin your home state." He waves his hand, idly gesturing to the land, the pump jacks, flares, and derricks. "The only thing left to do is move."

He's bought a house in Montana and invested in a ranch. He and his wife are building a business and a life there, but Bobby still has a few years before retirement. The past several days of work, he'd treated me strangely. Earlier that morning, he pulled the van around to tell me that he was proud of me. I laughed it off. "I'm proud of you, too, Bobby!" I hollered at him.

"What?" he said. "Me driving around in the van?"

Our conversation home rambles with no fixed destination. We talk about music, and he mentions his guitar. "You play?" I ask him.

"I used to," he says, "but that was a lifetime ago. Doing this work, there's no time for anything else."

He tells me about his investments, taxes, saving for retirement. We talk about the weather, the ice and snowstorms attacking the country. I explain the route I'm taking back to New York, and he tells me to be careful. "It would be a shame if you wrecked on your way home after this time in the oil field." He smiles only a little.

"Jesus," I reply. "It would."

At the end of the drive, back in Diamondback's yard, he wishes me safe travels. "It was a real pleasure," Bobby Lee says. "I always had fun working with you, Magic." He pauses to put out his cigarette. "Except in Glendive!" he adds. His face twists into a broad grin and I watch as he's enveloped in a full-body guffaw.

"Especially in Glendive!" I say, and hop out of the van.

I DROP BY Champ's house to pick up a W-2 that was sent to his address. The bunk beds are still in the living room, and the flophouse is still a wreck. Champ comes downstairs to meet me. His perpetual sunburn has faded in the winter months. His red hair is cut short and slicked back. We sit at the kitchen table, and I tell him I'm leaving town.

"You did good, though? You made a lot of money, right?" he says.

I lie. "Yes."

I arrived in Williston in the springtime of 2013 with close to $3,000 cash. I would be leaving in winter of 2014 with less than $2,000. How did this happen?

I made $21.25 an hour as a swamper at Diamondback. I got a lot of overtime, but also had weeks where I didn't work. My average two-week paycheck was $1,762. So that in thirty-four weeks, I cleared just under $30,000. It's honestly hard for me to fathom.

"You're headed back east," Champ says. "There's nothing back there for me now. It's all gone. This is normal now." He continues, "The house is full. It's been full since you left. At this rate, I'll have the mortgage paid off in twelve months. Sixteen months, maybe. Then I'm taking a Sawzall to those fucking bunk beds."

One of Champ's new boarders walks through the kitchen. He asks if we want any snacks or coffee. The kindness of the offer is jarring, and I look at him a full moment before responding. "No thanks," I say. Politeness has grown foreign to me.

Champ talks about the housing market, talks about oil picking up. He tells me about the people who have come and gone through the flop. A guy who brought a TV film crew in one night when Champ was away. A Hutterite who fell in love with a stripper. A Ukrainian kid who asked if he could bring a girl over. Champ said sure, and when he got home there was a line of guys in the living room, waiting to go upstairs and fuck her, one after the other. "They were running a train on her," Champ tells me. "I told him you gotta get her outta here, man. She came down. She was sooo fucking drunk, man."

I take Champ in. He looks good. Fit and sober, a bare-knuckle striver. I'll never see him again. A few years later, I'll hear that not long after I last saw him, he started having trouble on the job. Showing up hungover. Not having his shit together. Dropping his trailer in the wrong spot one too many times. Diamondback let him go.

When I leave, the guy who'd offered me coffee and snacks is

lying on the top bunk above the bed I used to sleep in. He says goodbye as I walk to the door. It stops me again, this second unprompted kindness, and I turn to look at him.

"Take care," I say.

"I will. I'm so happy," he says to me.

"You are?" I ask.

"I'm just glad to not be picking up trash on a construction site right now."

IN WARMER MONTHS, I had driven out to the Viking's place from time to time to drink beers and bullshit while he tinkered with his yard full of derelict John Deere tractors. We'd get properly lubricated and then ride around the field on ATVs.

"You gotta pin that motherfucker back, Magic Mike!" The Viking would excoriate me until I popped a wheelie across the full length of his yard. Then he'd reveal his stash of Tannerite, a binary exploding rifle target, dump a packet of aluminum dust into a jar of fertilizer, and hand me a .22 caliber rifle. I'd sight the target, pull the trigger, and blow the holy shit out of a pile of pumpkins.

"Wooo-hooo!" we'd shout.

The Viking also kept a trebuchet in his shed. "You would have a fuckin' catapult, wouldn't you?" I said the first time he wheeled it out.

"Catapult is spring loaded," he growled at me. "Trebuchet works with dead weight."

"Of course it does," I said. I'd hold the dog as he launched large rocks about a hundred yards into the surrounding fields.

On my last visit with him, the Viking takes me ice fishing. We drive out onto a frozen lake, bore holes into the crust, and, huddling inside a small tent, cast our lines down a single round hole. All day long, we pull perch and northern pike up from the depths of their icy wet habitats. We toss the pike back down the hole, but we keep the perch, frying them up and eating them that evening for dinner.

The following day, I accompany the Viking to look at an ice fishing mobile home he's thinking of buying. "That's the boom, I guess. Doesn't matter much, but when it's over we'll have nicer stuff." He shrugs. "I guess."

He walks me to my truck. "Take care, Magic Mike."

"You, too," I respond.

I JOIN THE Wildebeest in his bachelor's basement, where he holds court with his feet kicked up in an oversized recliner while I sit awkwardly on the couch. We drink Pendleton Whisky and talk about life in the patch. I tell him about my writing. I've decided I'm going to turn my journals into a book of some sort. When he accepts the notion with a thoughtful nod, I'm overcome with a sense of relief at an approval I didn't know I needed.

We'd become buddies after he trained me up. "I made you a good swamper and then they took you away from me," he complained. He was promoted to truck pusher when Olhouser retired, and he'd request me for jobs.

At his apartment, he tells me about the boom in the eighties, and I ask him what he did when it went bust. "I did a lot of poaching," he tells me. "But don't put that in your book." I ask him to explain further. He says, "You need money for some things . . . electric bills, stuff like that. But when it comes to feeding a family . . ." He trails off.

I thought it was funny that, of all the things he might not want me to put in a book, this was the one? Sure thing, big guy. But I was awed by the implication of his words. There is a finite amount of oil on the planet. If we don't have anything to replace it with, and soon, we need to contemplate a return to a preindustrial way of life. When the black gold is gone, the Beest is saying, be prepared to hunt.

As the evening dwindles, the Wildebeest's daughter drops by. A high-spirited seventeen-year-old, she dotes unabashedly on her father as they engage in a playful kind of bickering. For a short while his son had joined us in the oil field, working as a swamper. It is striking now, as it was then, how incredibly gentle he becomes in the presence of his children.

MY LAST NIGHT in town, Huck and I head out for steaks and beers. We walk into the restaurant and see Smash sitting at the bar by himself. He's dressed in designer jeans, a nice button-down shirt, and a flat-brimmed cap with a UFC logo across it. His goatee is trim and his skin isn't porous like it used to be, back when it was daily filled with dirt, exhaust, and invert. He's been out of the patch for a few months now, working at a locally run shop that provides oil field support services.

"Took me a while to get used to it," he tells us, "working eight to six. Indoors. Everybody is real nice. Nobody yells at you, calls you a dumb fuck. Boss is real laid-back." He shrugs.

"Pussy," I say.

Huck chuckles, and Smash takes a sip of beer. He then looks off in the middle distance and nods in acquiescence. "Pretty much," he says. "Work's all right, though. I'm still getting the hang of the machines."

Neither Huck nor I has seen him since he left Diamondback. We huddle at the bar, order our steaks and our beers, and talk about the time that has passed. Smash and I trade war stories about rigging for Charley. We talk about my first move, on the Sidewinder rig. And other moves: Nabors 104, and Precision 602, that one XTreme rig out down by Dickinson. We talk about Glendive, and the buffalo skins. We gossip about the other swampers and operators. We do our Olhouser impressions. It gets us bent over laughing.

Smash recovers first and wipes his face. "I wouldn't think I would," he tells us, "but I gotta admit, I fuckin' miss it."

"Aahhhh yeeeaaah," Huck and I reply with only a tinge of solemnity.

It's still early when we finish our meals. But Huck has to work in the morning, and I have a long drive ahead of me. We drink a toast to the Bakken and the Black Gold, to Diamondback Trucking, and to each other, three good hands.

"I'm gonna miss you, bud," Huck tells me in the parking lot.

"I'll miss you, too," I say.

CALL OF THE WILD

I leave North Dakota on February twenty-first. Not long after, Huck has his day in court. The charges, stemming from spitting on a police officer, carry a maximum possible sentence of five years' incarceration. The prosecutor originally asks for twelve months but the public defender works a plea deal, and Huck is sentenced to thirty days. He is ecstatic, and he vows to turn his life around.

"Only thirty days," he would tell me afterward, still vibrating with good fortune. And then, flipping on a dime like he does, "but I still went to jail."

Huck lit up describing the prison dorms to me: "Before the shitter wasn't covered, so you're just pinching a biscuit sitting there looking at all the inmates. Which is awkward. But they put a curtain up, which was great."

Huck took some books with him into the clink. He knows I love to read. So he's excited to tell me how he read Jack London's *Call of the Wild*. "I had this epiphany about getting beat down and coming out a hundred times better than before," he says. "The dog was just the straight badass taking on the wolf pack. It was the sickest book

I've read in a long time, since like *Where the Red Fern Grows*. Just some real shit."

"Who else was in there?" I ask.

"A kid who got caught with a bowl. A drunk driver who killed two people. There were some straight-up meth heads. And there was a kid in there that beat a dude to death."

"Really?"

"We didn't get along. But yeah, he beat the kid to death, I guess."

"He was awaiting trial?"

"He was going to prison for like twenty years."

"Damn."

Huck was sentenced to thirty days, but in the end, he served only six.

"I'm sitting there, I ate my Cheerios. I hear the door opening, all the latches opening and shutting, and the guard walks in and yells my name. I pop outta bed like a billy goat, and he's like, 'You're getting out.'

"'Are you fucking kidding me?' is what I said."

"He's like, 'Don't make me change my mind.'"

"Amazing," I say.

Huck continues, "They got my release date messed up, so they just let me go. I had four different release dates! I didn't say it but I was like, 'You all suck at your job.'

"My phone was shut off, but I had some money for commissary. So I took that and walked down to Simonton's and bought some chew. I put a dip in. Fattest dip I had in a long time. I got like a weird buzz off it. And I walked to my sister's house. Nobody was home but my mom's pickup was there. They always go to Walmart. So I start walking to Walmart, hoping I can catch them. And if not, I'm not even mad because I'm outside! And right when I walked into the parking lot, they were pulling in. My mom and my sister and her son. I ran over to them. It was this weird crazy reunion thing. It was pretty cool."

I can't help but ask, "They weren't like, 'Did you escape?'"

"My mom said that," Huck admits, "and my sister did, too."

He takes a long, cool drag off his cigarette, then looks up at me. "I will never be able to get into the 'I'm in jail' mentality," Huck says.

THE BEGINNING

I leave Williston and burn across the country in three days with the windows rolled up and the music turned loud. I don't have any trouble until I hit snow in the Pocono Mountains, a couple short hours outside of New York City. A tractor trailer nearly sideswipes me, and I'm forced to pull off to the side of the road after almost losing control of my pickup.

I arrive in the city to spend time with a friend who lost her fiancé to cancer only a few weeks before. I stay at her place for several days. She gets all sorts of mail and care packages from friends. One day a giant box full of Girl Scout cookies arrives at her apartment. We stuff our faces with them and talk about grief.

It is cold and sunny. I walk aimlessly around Brooklyn drinking iced coffee. I try to find an excuse to tell every person I meet that I just came from North Dakota, from the oil fields. I shove it into conversations where it doesn't belong. I feel a compulsion to talk about it.

One afternoon, I pass a young woman on the street holding a clipboard. "Do you have a minute to talk about fracking?" she asks.

I almost spit out my coffee. "Sure," I say.

She has dark hair and a round, untroubled face. "Fracking is the greatest ecological threat to the planet," she tells me.

"Why?" I ask.

"It is the greatest ecological threat to the planet," she repeats. She doesn't blink. She asks if I'd like to support the organization she works for. "We work with the governor's office to help shape and pass legislation to protect the planet," she says.

"So, it's a lobbyist group?" I ask her.

"No," she says. "We work with the governor's office and the legislature to create and pass legislation that protects the planet."

"Isn't that the definition of a lobbyist group?"

New York State would ban fracking in 2015. Meanwhile, the state's consumption of natural gas, produced by fracking, would continue to rise. This not-in-my-backyard brand of environmentalism results in a fracking boom in neighboring Pennsylvania, and raises worldwide emissions by adding miles to the trucking routes that feed New York's power plants.

Essentially, environmental activists in New York would protect their state's consumers from the destructive results of their consumption, all the while adding to the degradation of the planet— the moral equivalent, in my eyes, of voting to support a war and signing up the neighbor's kid to fight it.

AFTER A WEEK in New York, I drive south to Baltimore, where I plan to restart my life. I arrive frazzled, burned out and busted, feeling every bit of my thirty-eight years. I move the contents of my truck into a buddy's spare bedroom and go to sleep. I sleep and sleep and sleep. I sleep twelve and fourteen hours at a time, and when I don't sleep, I read and watch movies in bed, compulsively. Then I sleep again.

In the seven years I lived in New York, the perpetual youth machine kept me preoccupied. I hardly noticed that my Baltimore friends were buying houses, getting married, having children, and growing into lined faces. It happened gradually, but when I return to Maryland it stuns me. I arrive at a dinner party and my clothes feel the wrong size. I get brunch with couples and their kids and my mouth feels the wrong size. I can't say the right thing. Or speak like a normal person. My volume knob is broken. I seem only to holler when saying something inappropriate or mumble incomprehensibly when trying to explain myself. I feel the coil centered in my belly winding tighter. The easier the conversation, the tighter that coil seems to wind. I find myself looking at my friends for weaknesses, getting angry at the smallest perceived slights. I challenge lifelong confidants in tight-lipped arguments over truly trivial matters.

I can't find a job. I apply for bartending jobs and construction work, mostly. I had felt for certain that, if the anarchy of my résumé wasn't enough to entice an employer to hire me, the simple fact that I worked the rigs for nine months would be enough to at least get me an interview. I am one of the hardest-working motherfuckers on the planet! I also have a few good stories up my sleeve. The reality is that there are barely any jobs available in Baltimore. Of the jobs that are hiring, when it comes to my oil field work, nobody gives a fuck.

I land a job as a waiter in a trendy part of town. One bartender is called "Cage" because he looks and sounds like the actor Nicolas Cage when he styled his hair like Elvis. Cage keeps a comb behind the bar and sometimes, when attractive women order drinks, he makes eye contact with them while running the comb through his hair with a serious expression on his face. My own hair has grown long, it goes red in the summer, and I wear it wild.

The bar has a high turnover rate. People come and go every week, so I'm able to get lots of hours. I work double after double, staying on my feet day after day through twelve-hour shifts. The waste I see in the restaurant is shocking. I throw away more food than I see eaten. I come to think of my job as throwing away food. I see it outside of work as well. My friends and I consume expensive meals, buy new clothes, new gadgets, and new furniture. We fill recycling bins with plastic bottles, order packages from Amazon, and fly halfway across the world on vacations. As disempowered as cogs in an earth-eating machine, we seem to have lost the ability to say, "No, thanks." I begin to see this not so much as a political issue—I vote once every two years—but more as a problem of the culture. Why don't we stop buying all this shit?

I quit the job and I go on tour with a friend's band for a couple weeks. We stay stoned and play tunes and go swimming and ride in a van.

I borrow money. I drive to Colorado for a wedding, then out to Utah for a month of solitude in Park City, then to Los Angeles to visit Matthew and his daughter, and back through Colorado, up into Wisconsin, and finally back to Baltimore in late autumn of 2014.

The whole time I am restless, too tired or too awake, horny as hell, aimless and half-wrecked. I think constantly about returning to the oil field. I miss the guys. I miss the work. I miss the sense of purpose. I text with Huck, the Viking, Charley, and the Wildebeest, but they feel far away. I feel far away. I am in two places at once. I can't seem to fit in anywhere.

It surprises me. I thought if I could make it through the black gold rush, become a good hand, I could make it anywhere. I could do anything. Of course, I am wrong. Nothing I learned seems to translate in the slightest.

IN 2015, A young black man named Freddy Gray is killed in police custody, and the city of Baltimore erupts in mayhem. The National Guard is called in, and I march with protesters through the
streets. I realize that in some ways my shock at the racism I perceived in North Dakota was a reaction to its blatancy. It wasn't as
polite, or whispered in hushed tones, or denied in the way I was
used to it being denied.

I begin to dig at the embarrassment this revelation awakens in
me, trying to get to the root of it. At the same time, I am working
to hash out the incredibly difficult relationships I have with my
brothers and sisters. I begin to see a reflection of my country in my
own broken family. When a father abuses some of his children, he
creates a distinct, isolated reality for each of them. I grew up in a
big family, and yet in so many ways, we each grew up alone. We
entered adulthood without developing the proper tools to deal with
our own pain, much less each other's. Marching through the streets
of Baltimore is exhilarating, but it isn't until I begin to think of
black Americans as my own brothers and sisters that I'm able to
start to process the racism in my own heart, born of my own com-
plicity, willful ignorance, and shame. In many ways, this directly
mirrors my attempts to heal the relationships I have with my siblings.

OVER THE YEARS, Jack Jackson calls me regularly. He's in Florida when a car he's riding in is pulled over for a broken taillight. He
makes a run for it, somehow evading the police by racing down an
off-ramp. Then he's in Mississippi staying in a trailer his son owns,

fucking his son's girlfriend's mother, working on cars in the trailer park for extra cash. Or he's in Michigan, where a different son throws him out, and he moves in with a woman who hits him in the head with an ashtray, but he can't afford to get stitches so he's superglued his head back together. As always, he's funny and spirited, irascible and goofy. But as time passes, life on the run starts to wear him down. As a felon, he can't get a straight job, and the humiliation of endless, enduring poverty makes him bitter. At one point, he tries to turn himself into the police in Washington State to finish his time, and the bureaucracy turns him away.

"I seriously can't even get fucking arrested," he tells me. "If it were raining pussy, I'd get hit with a dick."

His phone calls become exhausting. The litany of bad news and busted hopes, the tightening noose of bad luck, bad circumstance. I quit answering his calls and one day they stop. Huck will tell me that Jack gets arrested for stealing equipment from a federal park. "He's not getting out anytime soon," he says.

North Dakota goes bust. Oil prices plummet from over $100 a barrel to just over $30. After I leave town in 2014, the more than two hundred active wells drilling in North Dakota dwindle to as low as twenty-four by 2016. Migration out of Williston starts as a trickle and turns into a reverse flood until not only the roughnecks and roustabouts are leaving—dropping the keys to their two-year-old pickup trucks off at the dealership where they leased them and skipping town in the beaters they drove in on—but also the folks who came to offer support services: the waitresses, bartenders, cab drivers, gas station attendants, and delivery drivers. By the time prices begin to climb again, a couple years later, the industry will have all but abandoned the Bakken, moving most major drilling operations to the Permian Basin in Texas.

. . .

IN THE SUMMER of 2017, the Wildebeest's rusty blood finally gets the best of him. When he dies, I fly back to North Dakota to pay my respects. Looking out the window of the plane as it approaches the landing field, my stomach drops. That familiar feeling of fear I experienced the first day of every rig move floods back through my body. I push past it, meet up with Huck, and head to the service.

At the memorial service, the funeral director opens up the floor for stories, and I rack my head for a Wildebeest tale that would be appropriate in a room like this. I can't think of a single story I can tell in a room with women and children. When I look around, I see groups of oil field hands fidgeting in the pews with their foreheads scrunched up. Neither can they. Afterward, we have a pretty good laugh about it. We cuss our way through those stories at the bar.

Huck and I roar through Williston for a few days. I see Charley, the Gruff Crane Operator, and some others. We swap stories and bad jokes. Williston has changed dramatically in a few short years. Both the boom and the bust seem to have come and gone. The traffic on Route 2 has settled down. And when we head to a bar, I'm surprised to see a mix of men and women, hanging out, laughing and dancing. Most of the guys I'm with have the day after the memorial off from work, so we get to actually spend some time together. Williston is a small, rural town again, like I'd always heard it was in those years before I arrived.

Huck is operating a crane at Diamondback, but he is ready to move on. "I kinda just wanna go to like a beach town or something and wash dishes," he tells me. "Just do something that doesn't define me like out here doing oil field shit."

Before I head back to the East Coast, I make Huck promise to visit me before Christmas. "We'll head up to New York, I'll show you around. You've never seen anything like it," I tell him. The thought of Huck wandering through the canyons of Lower Manhattan delights me.

A few weeks later, Huck posts a photo of his Harley-Davidson parked by the side of the road against the backdrop of a brilliant North Dakota sunset. Two days later, Charley calls to tell me that Huck had been out drinking with a new group of friends. He spent the night on a recliner in their living room and when they went to wake him in the morning, they found him dead. I'd later find out that he had mixed some form of opiates with hard liquor. It caused his lungs to start bleeding, a kind of instant pneumonia. And it killed him. He was twenty-five years old.

I go off by myself for several days of camping, walking trails in the Catoctin Mountains, listening for Huck's laugh in the tumbling stream. There is no getting over it. It breaks my heart then. It breaks my heart now.

I'd always imagined Huck and me in our old age drinking lemonade on a front porch somewhere, swapping reminiscences of our Bakken days as rig up hands, and laughing, always laughing, laughing lines in each other's faces, always making each other laugh more than anybody else ever. When I think of him now, I find my shoes untied, my keys misplaced, I forget what I was doing, and I look at the clock to see time has somehow passed never to return. I've got an empty shotgun shell and a picture of Huck on my desk. He's in the driver's seat of his truck, a ragged scar on his forehead from a motorcycle wreck, and a wide grin wrapped around his face. My brother. Fuck, I miss you, buddy.

. . . .

I CAME BACK from the oil field with a chip on my shoulder, a drinking problem, and a feeling that I had somehow figured it all out. I house a six-pack plus two or three every night for weeks, then months and then years—a river of amber pouring down my throat and out my piss. I fumble love, stumbling in and out of sex and relationships, drunken bull-in-a-china-shop-style. The wreckage piles up, and the only constant is me.

In Baltimore, I fall for a redhead with a broken nose and tattoos that cross her body like train tracks. Sara was homeless as a kid, and her life is defined by unrelenting hunger. We have that in common, along with a lust for tearing up the town. We start out as drinking buddies and, ignoring the advice of most of our friends, begin dating.

In some ways, our relationship is a continuation of the relationships I had with guys in the oil field. Sara and I are best friends, fiercely loyal to each other; we have a fuck-the-world attitude, and we are protective of each other. But our love is a rough love, inarticulate somehow, broken, built on scars. But Sara has a wicked sense of humor and an incredible knack for words, and I admire that in her.

When I was still in North Dakota, I'd once told Champ I was going to write a book.

"You're a thirty-seven-year-old laborer, dude," he said. Classic Champ. He wasn't wrong, but I wrote everything down, anyway.

Now I'm past forty. I start putting my words together again. I give Sara my pages and she marks them up, giving me insightful feedback and plain good advice. When she doesn't like something, she looks at me with a silly, baleful expression in her big brown eyes and moves her hand across the page in a scissors motion. Her cuts are always spot-on.

In tribute to her expressive eyes, her childlike enthusiasm, and desire to please, I call her my Puppy. We spend our nights drinking wine and smoking weed, making up ridiculous dance routines to Tom Petty songs. But Sara has wild mood swings, and sometimes she just won't read the pages. I see them sitting untouched by the side of the bed. I try to goad her into a goofy dance; she snaps at me. She grows withdrawn and depressed. She has a temper, and I never know what will set it off. Nothing seems to fill the deep well of need inside her. I twist myself into knots trying to. But when I myself don't know what I need, how can I help her? I am hapless and she is explosive. The day we move in together, she tells me to get the fuck out of her house.

Before I left North Dakota, I looked in the mirror and saw a good hand, but now I feel like a loose collection of bad habits. I'm still chasing thrills, getting drunk and getting stoned, then dragging my ass around, hungover all the time. I never have enough money. I can't seem to collect my thoughts. I'm a bundle of the same anxieties and insecurities that have dogged me since childhood, from the farmhouse with the hidden gun to New York to Williston and back again. Where does any of this end? When I look in the mirror I don't see a good hand. I see some drunken jackass.

Combat veterans in the American military have higher rates of PTSD than any of our Western first world counterparts, double that of the British, for example. Why is that? Because the United States does not rely on a conscription service, our military is populated primarily with men and women from the poorest, roughest parts of the country, or by those whose parents served. Just like the oil fields. No research has been done on field hands, but military men are twice as likely to report instances of sexual abuse as children than men who never serve. Think about the implications of that. These veterans have PTSD *before* they experienced combat. They leave home broken and come home more broken.

One afternoon, after weeks of continuous fighting, Sara and I are in couples therapy. We've been together nearly three years and it clearly isn't working. I goad her into a fight. She storms out of the meeting.

My college mentor passes away—the great Irish director Sam McCready, founder of Belfast's Lyric Theatre. I sit by his bedside in a cottage on the outskirts of Baltimore a few days before he dies, and he tells me about the play he has just finished writing. Our relationship, after college, had always been a little awkward. I think he wanted to be more of a father figure to me, but I rejected that. I was too independent, and he had what I viewed as a bit of a puritan streak in him. I thought he disapproved of my hedonism, and I didn't want to hear it. But we had made incredible work together. He was a bracingly good theater director. Sitting next to him, I can't get over how unashamed he is of being an artist. He has none of my baggage, none of my desire to be a mechanic or whatever the hell it is I think I should be. He knew I had been writing about the oil field. When I called to schedule a time to see him, he didn't say hello, he took the phone and said in his wonderful baritone brogue, "You must continue writing, Michael. You must! It is very important. You must keep writing, Michael! Now, when can you come see me?"

I tell this to my therapist, a wonderful Canadian doctor named Sandy who forces doughnuts on me every session. She wants to eat them herself but she won't admit it, and I adore her for it. She tells me that I deserve to be happy. I find myself crying. "Does this happen to all the men you see?" I ask her.

"It's a safe place," she says.

"I feel so guilty," I tell her.

"About what?" she asks.

"Sara," I say.

"Did you do something wrong to Sara?" she asks.

"No," I say.

"Then that's not why you feel guilty," she tells me definitively. "So, you need to figure that out."

I'd been trying to figure it out for weeks, for years, really, my whole life. But this would be the first time I'd ever say it out loud. "I knew my dad was molesting my sister," I tell her. "I was just a kid, but I knew. She asked me to come into the room. I can still see her face."

I'm in my bedroom at the farmhouse. Dad has been playing with us, tickling, squeezing, and pinching—his hands too rough even when it's supposed to be a game. He summons Kate to her bedroom, and after they leave she rushes back into my room, and asks me to join them. Standing in the doorway, she tries to act silly, but she's unable to hide a pleading fear that pulls and tugs at the contours of her face. "No," I tell her, and her façade falls completely. I know exactly what is going on. She looks at me again. No longer pleading. She is scared and resigned. I try to smile. "I'm going to bed," I tell her. She looks at the floor, then leaves the room in silence.

I tell this to Sandy, my face twisting in a knot, tears splashing into my hands. "You were just a child," she tells me. "What your father did to your sister was wrong. What he did to you was wrong. You were a victim, too. Look at me." She holds me with her eyes as I weep. "You're a good man," she says. And a part of me, a small part buried somewhere deep inside of me, not only hears her, but believes her, too.

Sandy tells me that I'm not responsible for what I'd been subjected to growing up, but I am responsible for what I do next. I haven't had a steady job for some time. It is 2019; I'm picking up days at my buddy's moving company—forty-three years old, carrying mattresses and dressers up three flights of stairs, then coming home to the ragged remains of a broken relationship. I know that I need to leave Sara, but I have nowhere to go. When she storms out of our last therapy session, I race home, pack a bag, and flee town. I move in with my mom.

THE FARM IS long gone. It was sold off years before. My mother now lives in a sun-filled row home in the historic district of a small town in western Maryland. She'd reinvented herself since the divorce from my father, going to college when she was in her forties and earning a degree in social work. She became a case manager at Child Welfare Services in Harford County and worked there for over twenty years. She has been instrumental in mending the lives and families of hundreds of parents and children.

I'm in the spare bedroom of her house drinking loads of coffee. For the time being, I've quit alcohol. I'll see how it goes. I've got papers all around me, copies of emails I sent when I was in the field, old journals still smeared with remnants of Williston dust, receipts, and swamper time sheets from Diamondback. On my computer are photos of the guys—Huck smoking a cigarette and trying to look cool, Jack Jackson bent over the hood of a car, turning a wrench, the Viking walking alone through prairie grass, his hands extended, palms touching the very tops of the leaves. I scroll through files of recordings I made talking to the guys, goofing off, drinking beer, and telling stories. I pull a blank page up on my screen, and I start to fill it.

Some days it's easy. Today, the words just spill out of my fingers and onto the page. I watch them splash across it. I'm back in the field. It's my first day, and my coveralls are too big for me. I haven't gotten used to the feel of the netting inside my green hard hat. I'm fiddling with it when some guy asks me if I've ever been turtle fucked. I'm trying to catch it all, trying to remember it exactly like it happened, trying to express it in a way that other people might understand, might take something from, might recognize. I want to understand it all myself. It is arrayed all around me like a thousand-piece jigsaw puzzle. I've never done this before.

As I'm writing, I'm not just recording, I'm creating myself. Putting the pieces of myself together. It isn't clear to me yet, but somehow, in some way, the insecure theater kid is the oil field hand, the hard-drinking guitar slinger is the sober guy at the desk. The coil at the center of his soul can be used for something besides self-destruction. It's a gift, this pain. It has to be. And he has to give it away, because holding it will kill him. Healing is the new job.

There's a half-eaten sandwich on the desk next to my computer. My mom made it for me, two pieces of white bread, ham, and mayonnaise. I'm embarrassed and comforted by its presence. I take a bite and chew slowly. Who puts mayonnaise on ham? My mom does, that's who. I stare idly at the floral design on the plate she left when she brought the sandwich up to me. It's hard to feel like a tough guy in these circumstances.

I've spent enough time with tough guys to know that they're actually just marshmallows. They've learned to push down their gentleness, to cover it with hardness and anger because they've been born in hard, angry places to do hard, angry jobs. When Charley attended Huck's funeral, he told me that he'd underestimated Huck. "Everybody got up and spoke about him as this gentle giant," Charley said. "How he'd touched their lives."

To see Charley with his wife was to witness a man gobsmacked by love. His kids climbed over him like puppies over a cranky old dog, and he adored it. To sit in his home, or in the apartment of the Gruff Crane Operator and his wife, or to visit the Wildebeest at his place, was to enter into what for me was a foreign land: a world attended to by nurturing, loving fathers.

During the darkest days I spent working under the Wildebeest, I had a dream that he was wearing a floppy straw hat, kneeling in the soil in a field of yellow flowers, gardening. It was the weirdest thing. You can't erase what a man is. But thinking of it months later, sitting at the desk at my mom's house, I wonder who he might have been in a kinder world.

There are no bookshelves in my room—all my furniture is still at the place I shared with Sara—so the floor around my desk is covered with books. *The Rise of Theodore Roosevelt*, *The Journals of Lewis and Clark*, *The Folk Songs of North America*, a collection of plays by Bertolt Brecht. I've got the short stories of Hemingway, Jack London, and Raymond Carver stacked next to thick, boring books on oil production, terrifying treatises on global warming, and several books on North Dakota's Native American tribes.

In one of those books, *Encounters at the Heart of the World*, I learn that during the Okipa ceremony, boys and men of the Mandan tribe would volunteer to participate in a ceremony that involved binding their penises with deer sinew, chopping off the little finger of their left hand, and suspending themselves from the beams of their lodges by leather tongs attached to wooden splints run through their chests and legs. The meaning of this, writes Elizabeth A. Fenn, "is easily misconstrued. Perseverance through pain was neither a rite of passage nor an emblem of manhood. It was an offering, a praise song . . ."

Ramblin' Jack Elliott is howling "Old Blue" through the speakers on my desk. I'm typing, shuffling through papers, picking up and putting down books, trying to make sense of this sprawling story that began as a story about an oil boom but has become something else, what exactly I don't yet know.

When I got my first role in a school play, my mom drove me to rehearsals every night in a beat-up station wagon with a broken heater and front windows that wouldn't roll up. It was autumn and we'd cover ourselves with quilts and blankets, often getting so cold that we'd burst out laughing. We put plastic on the windows, but the wind blew through them. At school, Mom dropped me off around the corner from the entrance so no one would see me climb out of the dilapidated car. It was held together by duct tape and the doors didn't work, so I had to crawl out the window. Mom never complained. She knew how important the school play was to me.

I close my computer, stand up and stretch. I take my coffee, the plate and remnants of my sandwich and walk downstairs to the kitchen. Mom is standing at the sink doing some dishes, humming along to the radio. I make sure she doesn't see me throw the uneaten half sandwich in the trash. I hand her the empty plate.

"How's it going, darling?" she asks me.

"Good," I say, heading out the back screen door of the house to sit in the sun.

I sip my coffee and let the words fall out of my head. Through the door, I see the silhouette of my mom still standing at the kitchen counter. Her head is bent over the sink as she scrubs my plate. I think for a moment about Huck—Mom would have loved Huck;

she's a sucker for a rogue. I think about the Beest—Mom would not have liked him so much. I think of Dad. I watch Mom as she sets the plate on the counter. From where I'm sitting, looking in, it is dark in the house, but for a second the light catches in such a way that I get a glimpse of her expression. Her face is lined from worry and heavy with time, but she looks peaceful.

She steps outside holding a coffee mug. "My rhododendrons are starting to bloom," she says, walking over to the plants. "Look at this." I move with her over to the bushes, and she fiddles with them as I watch. At the center of groups of large, flat green leaves sit small, tightly huddled buds just barely opening, revealing flashes of a light, delicate purple. "Maybe these are azaleas," she says. "I can never remember the difference. Who cares? They're pretty, though, aren't they?"

I'm not paying much attention to what she's saying. My mind is over a thousand miles and five years away. Mom is still bent over the flowering bush, but she notices I haven't responded. When she looks up, she appraises me with a critical eye. "What are you doing standing around staring at bushes with me, anyway?" she says, nodding back toward the house. I nod silently, follow her cue. Inside, I refill my coffee from the pot by the stove and walk back up the stairs to my desk. I get to work.

ACKNOWLEDGMENTS

First and foremost, I thank my mom and my siblings to whom this work is dedicated, not only for their incredible love and support over the years, but also for their willingness to talk to me about some very difficult memories, and their permission to write truthfully about some of our toughest days.

I also want to thank my dad, who passed away in late 2019. When I told him I was writing this book, he said, "Just don't write about your family! Skip all that! If somebody asks, say, "Childhood? What childhood?'" For that, from him, I remain eternally, hilariously grateful.

I am indebted to all of the friends and former coworkers I met in North Dakota, especially Tyler, Tibor, and Ashley; Chad and Karen; Kevin C., Kevin N., Adam, Matt, Big Dave, Casey, Kory, Kirby, Darren, Todd, Lloyd, Tanner and Taylor; Doyle, Nick, Chris B., Terry, Tom W., Tom L.; as well as Gibby, Beth, Craig, Chris, Blake, and Andrew, Scully, and every other hand I met and worked with along the way.

To Bill Clegg, I frankly have no idea how to express my gratitude. I'm heartened to know we are just getting started. The staff at the Clegg Agency are all champions: Simon Toop and David Kambhu were especially instrumental in shepherding this book to completion. Henry . . . gave me notes one key at a time.

I am incredibly grateful to my editor, Paul Slovak, not just for his gentle and meticulous editing, but also for the Ken Kesey stories! I want to thank Andrea Schulz for giving me this shot, and Lindsay Prevette at Viking, as well as Paul's assistant, Allie Merola.

Erin Whitenak's belief in my talent buoyed me over long stretches of time, even when my own belief in the project wavered. Erin's incredibly keen intelligence, passionate love of language, and devilish wit influenced many of this book's pages. I am incredibly grateful for her help.

I was itinerant during much of the composition of this book, and I'm thankful to the many people who housed me and provided me with places to write when I needed them. Jim Cricchi and Susan Peters also offered invaluable feedback and emotional support. You guys share the MVP award! Thanks also to Caleb Stine, Travis Kitchens, Kaityn McHugh and the entire McHugh family, Bucky Hayes, and Amy Baumgarten.

There were times when my desire to write and my desire to eat seemed to be at odds with each other; a few different folks hooked me up with work during a particular rough patch: thanks to Danny Balsamo at Making Moves, Justin Loys at Junk Dog, Cate at Balloons, and Adam Collison.

Several folks read this book as a work in progress and gave feedback. Thanks to Justin Mills, Chris Pumphrey, Elizabeth Evitts Dickinson, Ben Forstenzer, Zach Pontz, and Jamie Seerman. Dr. Sandy Snow offered me encouragement during a crucial period and also encouraged me to find the strength to deal head-on with some of the more difficult passages in this book. My acting mentors, Carl Freundel and Sam McCready, receive shout-outs in the pages, but I also feel compelled to thank them here, along with Natalie Rebetsky, my eleventh-grade English teacher and one of the first people to encourage me to write.

Many of my NYC and Baltimore friends read my emails while I was in North Dakota and encouraged me to put these stories into a book. Thanks! These folks include Kyle and Brigid Riley, John Atzberger, Mike Eisner, Shannon Cassidy, Megan Hamilton, Sarah Murphy, Jason Planitzer, Eline Gordts, Ann Courtney, Tommy Harron, Kristen Toedtman, Brendan Hines, Michael Beresh, Alan Kreizenbeck, Jamie Reeder, Ben Jaegar-Thomas, Jacob Thomas, Trixie Little, Bryan Dunn, and Emily Raw.

Although much of the history of Pithole did not make it into the final version of this book, many books about the world's first oil boom added greatly to my understanding of what I experienced in the Bakken. Chief among them are the following: *Petrolia: The Landscape of America's First Oil Boom* by Brian Black, *The Great Oildorado* by Hildegarde Dolson, *Pithole: The Vanished City* by William Culp Darrah, *Sketches in Crude Oil* by John J. McLaurin, and *The History of Pithole* by Crocus.

Two books by Sebastian Junger supplied me with a lens through which to view my experience with the men I worked with. I read *War* while I was working the rigs, and I read *Tribe* after I'd returned to Baltimore. Both works profoundly affected my thinking.

The quotes from Woody Guthrie used as epigraphs in the first and third parts of this book come from *Pastures of Plenty: The Unpublished Writings*, edited by Dave Marsh and Harold Leventhal. The Charles Bukowski poem quoted in the Rosalie chapter is called "Love Poem to a Stripper" and can be found in *You Get So Alone at Times That It Just Makes Sense*. The Waylon Jennings lyrics in the Boomtown Earth section come from the song "Are You Sure Hank Done It This Way." The poem, "God's Grandeur," at the beginning of the book is from Gerard Manley Hopkins.

Other helpful books include *The Prize: The Epic Quest for Oil, Money & Power* by Daniel Yergin, *A Brief History of Williams County, North Dakota* by Ben Innis. *The New Wild West* by Blair Brody, *The*

History of North Dakota by Elwyn B. Robinson. Information on Lewis and Clark primarily comes from *A Vast and Open Plain: The Writings of the Lewis and Clark Expedition in North Dakota, 1804–1806*, but I also consulted Stephen Ambrose's *Undaunted Courage*. Information on the tribes of North Dakota comes from *Coyote Warrior* by Paul VanDevelder, *The American West* by Dee Brown, and *Encounters at the Heart of the World* by Elizabeth Fenn. The chapter on Theodore Roosevelt owes its existence to several sources, chiefly Edmund Morris's unimpeachable *The Rise of Theodore Roosevelt*.

My interest in the North Dakota oil boom was first piqued by an article in *Men's Journal* called "Greetings from Williston, North Dakota," written by Stephen Rodrick. Articles by Susan Elizabeth Shepard in *Buzzfeed* and Laura Gottesdiener in *Mother Jones* were also sources of inspiration for me when I began writing.